RED MITES AND TYPHUS

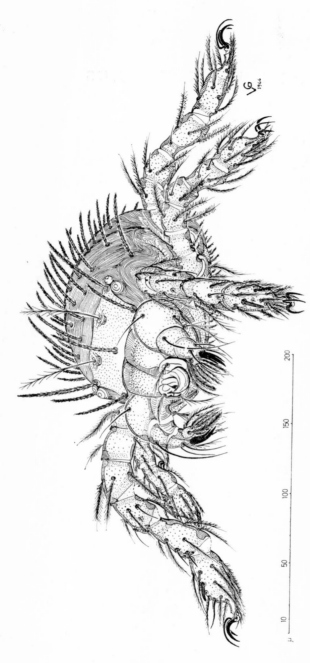

FIG. 1. *Leptotrombidium* (*L.*) *akamushi* (Brumpt), widespread vector of scrub-typhus. Example of an unengorged chigger or larval trombiculid mite. Most drawings of chiggers are from flattened specimens mounted on slides, but this exceptional drawing gives an impression of the live larva. It would become greatly distended after 1–3 days of engorgement on its host. Drawn by, and reproduced by the courtesy of, Dr P. H. Vercammen-Grandjean, The G. W. Hooper Foundation.

UNIVERSITY OF LONDON
HEATH CLARK LECTURES 1965
delivered at
The London School of Hygiene and Tropical Medicine

Red Mites and Typhus

by

J. R. AUDY

The George Williams Hooper Foundation
San Francisco Medical Center
University of California

UNIVERSITY OF LONDON
THE ATHLONE PRESS
1968

Published by
THE ATHLONE PRESS
UNIVERSITY OF LONDON
At 2 Gower Street, London WC1
Distributed by Constable & Co Ltd
12 Orange Street, London WC2

Canada
Oxford University Press
Toronto

U.S.A.
Oxford University Press Inc
New York

© *University of London,* 1968

485 26318 1

Printed in Great Britain by
WESTERN PRINTING SERVICES LTD
BRISTOL

This small part of a long story is dedicated to the many hundreds who have died of scrub-typhus acquired in the course of their work as farmers, labourers, soldiers or investigators; also to leaders in research on scrub-typhus from three countries, namely Dr R. Lewthwaite, C.M.G., the late Dr Joseph E. Smadel, and the late Dr Takeo Tamiya, as well as to my colleagues in the 'Imphal Circus' and the research units in Myitkyina and Kuala Lumpur.

PREFACE

When I was invited to give these Lectures, I was told that I should not be too scientific or technical but that, observing the terms of the Heath Clark Trust, I should attempt to deal with aspects of history, education, and the relationship of the humanities to science. I have endeavoured to do just that in each lecture, praying that my musings will not give a disjointed impression. A solid scientific account would have been a much easier task, for I have at times suffered what may have been felt by the legal draftsman who drew up the terms of the Heath Clark Trust itself. He was evidently enjoined to restrict the honour of giving these Lectures by some sort of exclusive clause—but at the same time not to close the door too tightly. Presumably as a result of what the psychiatrists would call a double-bind, Article 8 of the Trust emerged thus: 'The Lecturer may be a teacher in or a member of the University of London or any other University whether within the British Empire or elsewhere or a person of either sex (whether a British Subject or not) who is not a member of or attached to any University.'

It is hard to be wise, nor do people take much heed of wisdom. It is much easier to be provocative—and people do heed provocation. Choosing the easier path, I have been deliberately provocative here and there, particularly in Chapters 1 and 2.

For publication, the Lectures have been left more or less in their uncondensed form. In addition, more references, some technical details, and more translated quotations, have been presented as appendices. This is in order to make the published Lectures more useful to the student, and also to quell the urge I have had for many years to write a book on scrub typhus that would have given at least some of these details. For thus making it unnecessary to write that book I am very grateful to the Heath Clark Trust and the Athlone Press, especially to Mr A. M. Wood and the late Mr W. D. Hogarth.

I deeply appreciate the invitation by Professor E. T. C. Spooner and Dr James Henderson to give these Lectures (the responsibility for which is heavy and sobering), and also the introduction of Dr R. Lewthwaite by which he further honoured me. For help with translations from the Japanese, I am indebted to Dr Nobuo Kumada, a visiting professor at the George Williams Hooper Foundation for Medical Research at the time I started to prepare these Lectures, to Dr C. B. Philip, and especially to Miss Naila Minai and Mrs Atsumi Minami. For translations from other languages I am grateful to Mrs Rosalie Watkins and Mrs Margot Mandel. For untiring effort, my greatest debt is to Mrs Iris McRae, Editorial Secretary, and to Miss Esther Picazo, Miss Avis Easter, Mrs Julie Treftz and Mrs Lucille Valentine, for seeing the manuscript through its final stages.

I am also grateful to Dr P. H. Vercammen-Grandjean for the original drawing, Fig. 1, and to Miss Marjorie Smith for drawing Figs. 5–10 from material in my possession; to the late Dr Takeo Tamiya, Dr Masami Kitaoka, and (the photographer of several) Dr Kiyoshi Asanuma, for the 35 mm slides from which Figs. 5–9 were copied, with their permission; for permission to reproduce figures, to Dr Kiyoshi Kawamura (Fig. 2), Dr C. B. Philip and the U.S. Army Medical Museum (Fig. 3), and Dr Jacques M. May (editor) and the Hafner Publishing Company (Fig. 4); to Dr Ungku Omar-Ahmad, Director, the Institute for Medical Research, Kuala Lumpur, for permission to reproduce the cartoon, Fig. 13, and, in Chapter 3, parts of my text on typhus from the Jubilee Volume, *The Institute for Medical Research 1900–1950* (*Stud. Inst. Med. Res. Malaya*, No. 25, 1951).

J.R.A.

CONTENTS

1. **SCRUB-ITCH AND THE ECOLOGIST**   1
   Introduction, 1
   Chiggers and Scrub-itch, 8
   Chiggers and Behaviour of Hosts: Ecological Labels, 12
   The Act of Feeding, 15
   The Observer versus the Experimentalist, 18
   The Ecological Outlook, 23
   References, 26

2. **AKAMUSHI: THE RED MITES OF JAPAN**   28
   Introduction, 28
   Earliest Accounts, 29
   The Tsutsugamushi, 38
   Early Recognitions of Transmission of Diseases by Arthropods, 40
   Development of the 'Miasma Hypothesis', 41
   Subsequent Scientific Investigations in Japan, 52
   Rivalry, Reason, and Research, 54
   References, 60

3. **EMERGENCE OF THE TYPHUS GROUP OF FEVERS**   63
   Introduction, 63
   First Clarification of Ideas: The Typhus Group, 67
   Typhus n Malaya, 69
   The Weil-Felix Reaction: Further Clarification, 70
   Two Kinds of Tropical Typhus: 'Shop' and 'Scrub', 72
   The Second Phase of Malayan Research, 73
   Evolution of the Typhus Group of Fevers, 76
   References, 82

4. **THE IMPHAL CIRCUS**   85
   Introduction, 85
   Peculiarities of the Ceylon Outbreak, 88
   The Concept of Animal Weeds, 91
   Jungle Tsutsugamushi and 'Man-made Maladies', 95
   The Scrub-Typhus Research Laboratory: The 'Circus' at Imphal, 97
   Controversy about the Vector, 102
   Doubts about Relations to Vegetation, 105
   References, 110

CONTENTS

5. **OLD AND NEW HORIZONS**   113
   Introduction, 113
   Changing Patterns of Disease, 113
   Changing Patterns: 'New' Diseases, 115
   Revision of Ideas about Rats, 117
   Jarak: Isle of Rats, 119
   References, 124

   APPENDICES   126
   1. Classification of Acarina, 126
   2. Scrub-itch: Some quotations from the literature, especially Oudemans, 1912, 128
   3. Pest-arthropods, with special reference to pest-trombiculids (chiggers), 143
   4. Japanese and other names for chiggers and scrub-typhus, 153
   5. Major publications on scrub-typhus and ecology of the vectors, 161
   6. A note on miasma and mosquitoes as causes of disease, 165
   7. Notes and translated excerpts from Japanese publications, 168
   8. Types of outbreaks of scrub-typhus, 172
   9. Mass flowering of bamboos and some consequences, 176
   10. Prevention and treatment, 178

   INDEX   181

PLATES

Figures 11 and 12 appear as Plates between pages 110 and 111.

I

SCRUB-ITCH AND THE ECOLOGIST

NEARLY two centuries ago the naturalist Gilbert White sat at his desk penning one of the many letters which were later to become a classical account of natural history in Britain: *The Natural History of Selborne*.[1] It was 30 March 1771 and the Reverend White started his 34th letter to Thomas Pennant, Esquire, with the following words:[2]

There is an insect with us, especially on chalky districts, which is very troublesome and teasing all the latter end of the summer, getting into people's skins, especially those of women and children, and raising tumours which itch intolerably. This animal (which we call an harvest-bug) is very minute, scarce discernible to the naked eye; of a bright scarlet colour, and of the genus of *Acarus*. They are to be met with in gardens on kidneybeans, or any legumens; but prevail only in the hot months of summer. Warreners, as some have assured me, are much infested by them on chalky downs; where these insects swarm sometimes to so infinite a degree as to discolour their nets, and to give them a reddish cast, while the men are so bitten as to be thrown into fevers.

Reverend cleric though he was, White could express himself vigorously, yet he dealt very mildly with a pest that has set back expeditions, kept people indoors and out of their infested gardens, and confined the naturalist and explorer Alfred Russell Wallace to his house for nearly two months while he recovered from a massive attack. The reason for White's mildness is probably that he wrote that letter in March when memories had been softened. Had he written it in August it might have read differently:

[1] Originally *The Natural History and Antiquities of Selborne*, published in 1788 (dated 1789), this comprises a series of wide-ranging letters, most of which were addressed to the Welsh zoologist Thomas Pennant and the antiquary-naturalist, the Hon Daines Barrington. As *The Natural History of Selborne*, these letters have been repeatedly published in a variety of editions, including paperbacks.

[2] White is obviously indebted to Henry Baker (1753), some of whose phrases he uses but without acknowledgement. Baker's quaint passage on the harvest-bug is reproduced in Appendix 2, which also quotes other accounts of these pests (chigger mites causing scrub-itch).

There is an insect with us, especially in my garden this summer, which swarms onto the innocent passer-by in its myriads, getting into the skin and raising lumps which itch so intolerably that the strongest man is reduced to the semblance of a whimpering child and may be thrown into a fever.

We now know the harvest-bug or mite—which the French call the *bête rouge* or *rouget*, the little red pest—to be the parasitic larval stage of a mite of the family Trombiculidae, subclass Acarina (Acari).[1]

The name for the family is apparently meant to suggest a resemblance to a diminutive thrombus or blood clot. The six-legged larvae of trombiculid mites are just visible as animated specks, usually red in colour. The larva takes a single feed of tissue-juices from one or a few particular species of animal hosts, climbing to a favourite feeding site such as the ears, and remaining attached there until it is engorged. This may take a few days, or even weeks or months, according to the species of mite and host and the prevailing temperature. The engorged larva then leaves the host in order to seek suitable shelter in the soil or the nest, where it pupates and later emerges as a very different four-legged creature covered in a dense pelage of hairs: a sexually immature nymph which later moults to become a sexually mature adult. The nymphs and adults are not parasitic but feed on eggs and various resting stages of insects and mites in the soil.

To illustrate this life-cycle, I have chosen not one but three charts, each telling the same story in a different way (Figs. 2–4). The charts happen to refer to scrub-typhus, a rickettsial disease, but the trombiculid mites that transmit this infection have a life-cycle which is characteristic of the Trombiculidae generally. The rickettsiae are virus-like organisms, visible under the microscope and characteristically associated with insects, ticks, and mites. One group of these organisms causes the various forms of typhus fevers, of which the best known is the notorious epidemic typhus or jail-fever, transmitted by lice.[1]

The first chart, from a Japanese report in Singapore in 1945, is delightfully drawn and very clear. It even conveys some

---

[1] See Appendix 1 for a note on the classification of the Acarina and Appendix 3 for references to trombiculids.

SCRUB-ITCH AND THE ECOLOGIST

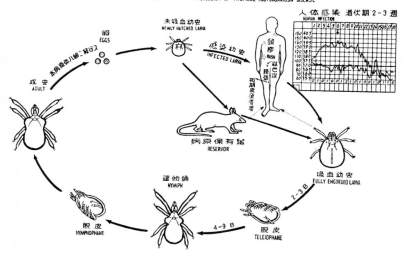

Fig. 2. Life-cycle of a trombiculid mite (chigger), I. Showing clinical details of tsutsugamushi disease or scrub-typhus. Reproduced with the permission of Dr Kiyoshi Hayakawa from a bulletin printed in Japanese from the Japanese Army Institute of Preventive Medicine (formerly King Edward VIII College of Medicine), Singapore, while he was Deputy Director: K. Hayakawa (1945). [A comparative study of Japanese and tropical tsutsugamushi disease (*R. orientalis* var. *tropica*) *Bull. Nanpogun Boekikyusui Bu, Singapore*, No. 100. (See also report summarized in *Trop. Dis. Bull.* 46, 788–790, 1949.)]

Translations are as follows – Between Adult and Eggs: *Pathogenic agent transmitted to eggs*. Above clinical chart: *Human infection. Incubation period 2–3 weeks*. On the figure, reading down and right to left: *Rash. Lymph node. Swelling. Initial skin-ulcer (eschar)*. The words in English, *nymphophane* and *teleiophane* (larval and nymphal pupae) should be transposed (the ideographs, *Moulting*, are the same for both). The legs and mouthparts of the larva do not increase in size during engorgement.

clinical details; but it shows man directly interposed in the life-cycle with an arrow continuing on from the man to the engorged larva. The rat, the host which maintains the trombiculids in the same way that rabbits, rodents, and birds maintain the harvest-mite in Europe, is shown almost as an incidental creature. This chart was designed by a clinician—or at least not by an epidemiologist—for it shows an anthropocentric attitude. An epidemiologist or parasitologist would have shown the rat in the ellipse with an arrow leading off to man as a 'dead-end' for the larva; for man is an accidental host. Larva$_e$

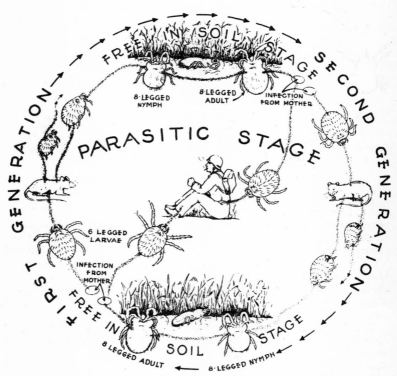

FIG. 3. Life-cycle of a trombiculid mite (chigger), II: 'Diagrammatic scheme of theoretical rat-mite-rat cycle of scrub typhus in nature. The "rickettsial stream" is continuous from generation to generation of chigger mites with new lines started from infected rats. Soldiers were accidental intruders in the cycle.' From C. B. Philip (1949). Scrub typhus or tsutsugamushi disease. *Scientific Monthly* **69**, 281–9. Reproduced by courtesy of Dr C. B. Philip and U.S. Army Medical Museum. The above caption however is from a reproduction (on p. 276) of the same chart by C. B. Philip (1964). Scrub Typhus and scrub itch. Ch. 11, pp. 275–347 in *Preventive Medicine in World War II*, Vol. VII. Washington, D.C., Office of the Surgeon General, Department of Army.

that attach to man have almost no chance of completing their feed and continuing the life-cycle. A pedant would note that the legs of the larva seem to have grown larger after engorgement: this is impossible, for only the body becomes distended.

The second chart is a textbook type of diagram correctly showing man as a dead-end. This chart shows the infecting organism as a sort of rickettsial stream of tiny dots, for the rickettsiae are transmitted through the ovaries of the female to a proportion of the eggs. Since the larva takes only one feed before it changes into the non-parasitic post-larval stages, it follows that it must already carry the infection when it takes its first meal on a man. It was therefore known with certainty that the rickettsia must be transmitted from the parent mite to at least some of the offspring—the technical term is transovarial transmission—long before anyone proved this experimentally. It simply had to be so. A pedant might criticize this chart by noting that it shows two stages, 'parasitic' and 'free-in-soil'. There is nothing to indicate the latter are not also parasitic, and that the larvae are 'free-in-the-soil' while they are waiting for a host and again while they are seeking shelter after engorgement. This chart demands closer study than the Japanese version.

In the first chart, the rat is labelled as the reservoir of the infection. Since the rickettsiae are passed transovarially from parent mite to offspring, there has been controversy, sometimes heated, as to whether the trombiculid or the vertebrate host is the true reservoir. Much of this controversy has been semantic. People have entered into this argument without first defining exactly what they mean by 'reservoir'.

The third chart is meant to be epidemiological. It devotes little space to the cycle in the soil but stresses the types of host and manner of distribution. A mite-colony will die out through natural causes unless it is reinforced by engorged larvae. James Gentry (personal communication) has found larvae continuing to appear in a colony of scrub-typhus vectors up to 30 weeks after exclusion of all animals. Hosts such as field-rats and quail have small home-ranges and they revisit the same sites frequently enough to return a proportion of larvae, now engorged, to the same colony or 'mite-island' (see page 11). The reliability

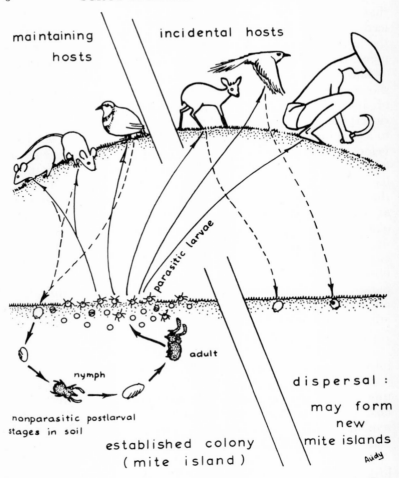

FIG. 4. Life-cycle of a trombiculid mite (chigger), III: 'Established colony (mite island). The diagrammatic representation at the lower left, of a chigger mite colony or mite island, indicates the chiggers' life cycle from engorged parasitic larvae through the nonparasitic postlarval stages (nymph and adult) to eggs and unfed larvae. The maintaining hosts are those that will reliably and regularly return a proportion of the engorged larvae to the mite colony, thus offsetting the losses due to mortality and transport away from the colony by incidental hosts.' This chart is intended to be epidemiological and to distinguish those hosts (M) responsible for year-to-year maintenance or boosting of a parasite – whether it be chigger, tick, sandfly,

and frequency of such visits are a requirement decided by the very short feeding-time of the larvae, a matter of two or three days. Thus, by their behaviour these hosts maintain or even boost the colony over the years. Another way of expressing this is that the life-cycle of the mite is so adapted to the range and regular behaviour of the host that populous mite-colonies can be built up rapidly. Other hosts, such as the mouse-deer and the crow-pheasant shown in the chart, are more wide-ranging and carry the larvae to places farther afield. By the time larvae have completed their brief feed, such hosts have usually wandered far away. These latter are incidental hosts and their main function is dispersal. The two types of hosts act together to establish the mites over the countryside. Man is a dead-end incidental host. These particular trombiculids shown in the charts have a wide host-range, but many species have definite host preferences; for example, some will infest birds but not mammals.

I would recommend to the student of parasites that he study again the many different life-cycle charts in various publications to discover the main purpose behind the design of each chart and to detect what biases the designer may have had. A diagram may be eminently suited to the original publication, but its reproduction may be inappropriate in some other work, for which a different kind of diagram would have been much more suitable.

The larvae of trombiculids are almost exclusively parasitic on vertebrates and especially on mammals. Largely through the adaptation of the larvae to feeding on vertebrates, this family has evolved from the Trombidiidae, the larvae of which

---

mosquito, or microorganism, virus – from those (I) that distribute the parasite or have no particular role in its existence. Maintaining hosts M include animal *reservoirs* of infections, but also hosts that maintain parasitic vectors. Although a *zoonosis* is usually defined as an infection such as scrub-typhus naturally transmitted between man and animal, it could be more usefully defined as an infection picked from M by a susceptible host I that is either man himself or a domesticated animal of importance to him (livestock, pet). From J. R. Audy (1961) The ecology of scrub typhus, page 393. Ch. 12, pp. 389–432 in J. M. May, Ed., *Studies in Disease Ecology*, New York, Hafner. Reproduced by courtesy of Dr May and publishers.

usually feed on insects and also on reptiles.[2] Some adult trombidiids are scarlet giants the size of a pea or a grape. These conspicuous mites may occur in very large numbers and because of their dense pelage they are commonly known as velvet-mites.

Well over a thousand species of trombiculid mites are now known. Of these, some 15 species cause scrub-itch in man by the allergic reaction to their saliva. These species are mostly in the genera *Eutrombicula* and *Schoengastia* but the harvest-mite is *Neotrombicula autumnalis*. Some 6 species (in one genus, *Leptotrombidium*), transmit the rickettsia of scrub-typhus or tsutsugamushi disease to man, while these species and certain others (including species of *Ascoschoengastia*) harbour and transmit rickettsiae among their animal hosts. This infection is geographically confined to an area bounded by Japan, West Pakistan, Australasia, and sundry islands in the Indian and Pacific Oceans.

CHIGGERS AND SCRUB-ITCH

Every continent has one or more species of trombiculid mites that have a very wide host range, including man, and that have saliva containing allergenic substances to which man and other animals may be sensitive or may become sensitized; and it is this delayed reaction which produes the intolerably irritating bumps on the skin, in the centre of which the living larval mite may often be seen as a tiny scarlet speck partly embedded in the swelling tissues. This gives the impression that the mites burrow into the skin. The resulting rash in man is known as *scrub-itch* or *trombidiosis*.[1] In Latin America such little creatures which attack the skin are often known by some derivative of the Spanish *chico*, meaning small.[2] The name

---

[1] See Appendix 2 for quotations from the literature and Appendix 3 for a brief account of pest-arthropods including trombiculids.

[2] Hence, we have been told, the *chigoe* flea (*Tunga penetrans*) which, after repeated introductions from South America to West Africa, did not start on its dramatic and dreadful spread along African trade-routes until just a century after White wrote his letter about the harvest-mite.[3, 4, 5] However, Hoeppli[5] states that the name of this flea was *chique* in the West Indies and, later, Senegal, and that this name derives from an American Indian word, *sika*.

chigger was applied first of all to the larvae of trombiculids causing scrub-itch in America, then to larval trombiculids generally and finally to all stages of these mites. It appears that those chiggers that cause scrub-itch are primarily species that typically feed on reptiles or birds but accept a very wide range of hosts, including man. We surmise that the marked reaction in man may be associated with the fact that these are (or, in recent evolutionary history, were) bird- and reptile-chiggers feeding on mammalian hosts.[6] This hypothesis would repay further investigation.[7] Since birds evolved from the common reptile stem a very long time after the mammals separated from it, many bird parasites show very close affinities with reptile parasites, and this is so true of trombiculid mites and some other parasites that my colleagues and I in Malaya used to say 'Birds is reptiles!'

Only those reptile- and bird-chiggers with an exceptionally wide host-range will feed on man. This wide host-range gives the species a powerful advantage in survival. It is therefore no accident that the species that cause scrub-itch in man are locally very numerous and vigorous, so much so that they may be a serious occupational hazard to man, as several scientific and many military expeditions have discovered to their cost (see Appendix 2).

Some pestiferous arthropods occur in such large numbers locally that they are either known or suspected to influence the distribution of larger animals as well as the seasonal behaviour of man. For example, midges and mosquitoes affect seasonal movements of herds in the subarctic, and blackfly (*Simulium*) make many valleys in the Balkans uninhabitable by livestock. The building up of populations of soft-ticks (*Ornithodorus*) in caves and burrows may, it appears, finally drive out the inhabitants, and it seems that the same might sometimes apply to man. It is possible that huts may perforce be abandoned because of such vermin—many Africans have found that livestock kept in their huts tend to keep down the population of soft-ticks as well as to divert their attentions temporarily, so that a very popular device is to have a concentrically double-walled hut, between the two compartments of which the family alternates with its livestock, being driven from one to the other every few months by the large numbers of soft-ticks. A century or so ago, Turkmenian and Uzbek culprits were punished by being confined in a cell with ticks—indeed, the sentence was a measured quantity of ticks determined for each prisoner! This novel form of severe punishment, described by Petrishcheva, must have been devised by captors who learnt their lessons directly from the ticks in rock-shelters. We know that scrub-itch chiggers do influence man's behaviour to some extent, when he tries to avoid contact. It has been reported that some parts of the Mato Grosso in Brazil are almost unexplorable because of chiggers, probably of the species *Eutrombicula batatas*. Evidence should be sought for their influence on animals also. The bush-turkey (*Megapodius*) of Australasia builds up huge populations of pest-chiggers. Do these in turn affect the other creatures, as they certainly do man? (See Appendix 3.)

Scrub-itch chiggers and their habitats are always well-recognized by the local inhabitants, who give them names that refer to their seasonal incidence, their habitat, their effects, or their appearance (e.g., red bug, *bête rouge*, *rotmilbe*). The diagnosis of scrub-itch may be overlooked for a time, as with 'autumnal erythema' in Czechoslovakia.

Let us return to Gilbert White's observations on the harvest-mites. He remarks on their seasonal appearance in the late summer so that they would be a pest to harvesters, hence the name, harvest-mite. He does not ask where they go in the winter time.[1] We now know that, although a few larvae may be found throughout the winter, at least on rabbits, there is a peak of egg-laying and subsequent hatching of hungry larvae in the warm summer. This is characteristic of many trombiculid mites throughout the world. There are, however, winter species. For example, in the Oriental region many larvae of the genus *Leptotrombidium* appear in the hot, wet summer whilst those of the genus *Helenicula* appear in the cold, dry winter.

Gilbert White comments thrice on the distribution of the harvest-mite in his part of south England. They are most active over the chalky downs. There is a concentration of them in

---

[1] Had he asked, he might have found an answer wide of the mark. It was once seriously believed that swallows spent the winter rolled up in balls of mud at the bottoms of rivers.

vast numbers on rabbit warrens; and they are also found in gardens. (In Denmark, the harvest-mite is almost confined to gardens, and to one town—see below.) These mites would indeed give us a good lesson in zoogeography or, in the present instance, medical geography. The distribution of a trombiculid species tends to be on three different scales. The first is the broad geographic distribution of the species, which in the case of the European harvest-mite *Neotrombicula autumnalis* stretches from Britain through most of Europe. Within the geographic distribution of the species, the mites are either restricted to, or at least mostly heavily concentrated in, particular endemic regions, the chalky downs of south England being an example in the present instance. Finally, within these regions we find highly localized concentrations of mites to small patches which may be less than a yard in diameter. North America has at least five species of pest chiggers, and this is what Ewing says of the distribution of one of the commonest, *Eutrombicula alfreddugesi* (the *Trombicula irritans* of authors): In the more humid southern parts of its range the mite is found wherever there is rough growth of weeds and shrubbery. Towards the northern limits of its range the species occurs only in isolated 'islands' where the local conditions are favourable for its maintenance.[8] It was this passage that led me to use the terms *mite-islands* and *typhus-islands* during wartime work on scrub-typhus in Ceylon and Burma. On this small scale, however, two terms are really required. The smallest unit of distribution, is the *mite-colony* in a very small place where an engorged larva or a batch of engorged larvae have reached maturity and are producing the next generation of unfed larvae. As we have noted, the colony may be boosted when the animal host regularly returns to the colony site at intervals that are short compared with the usual feeding time of the larvae. With the more vigorous and numerous species of chiggers that live out in the open, a number of different individual hosts may contribute together in seeding out colonies which may be practically confluent. It is therefore convenient to reserve the term *mite-island* for any localized unit of chiggers, often larger than a colony, the terms *mite-island* and *mite-colony* becoming identical in the case of a single isolated colony. A species of chigger is therefore distributed in the following successive scales: *geographic range, endemic region, mite-island*, and *mite-colony* (the last two sometimes being identical).

Ewing's comments on the distribution of *E. alfreddugesi* in the northernmost limits of its range might be relevant to the curious restriction of *N. autumnalis* to the town of Thisted and four near-by villages in Jutland. The harvest-mite has hardly spread into the countryside, being practically confined to gardens in parts of the town. It was first recorded as a pest in 1874, and according to Tuxen it is now such a plague to the 10 000 inhabitants of Thisted that they are 'virtually forced to live indoors in August'.[9] Two to four hundred pustules 'may be the result of a fine Sunday' in one's garden. Indeed, this number could be picked up on one's white tennis shoes after a minute or two of walking on a lawn at the time of maximum chigger activity before dusk.

CHIGGERS AND BEHAVIOUR OF HOSTS: ECOLOGICAL LABELS

Gilbert White draws attention to another important feature of attack by chiggers, namely its tendency to be occupational. Warreners are liable to severe attacks because their work takes them to the heavily infested rabbit warrens. Similarly, berry-pickers may expose themselves to heavy attacks by North American pest-chiggers. The principle here is as important as it is simple, that sharing a habitat risks sharing of parasites. These include parasites such as pest-chiggers, which may directly cause trouble, as well as others that may transmit infection, for example, certain species of chiggers that carry the rickettsiae of scrub-typhus. A parasite must be maintained by some host and the hosts are distributed in patches of various sizes and densities. In addition to dependence on the host for food, a parasite frequently has its own environmental requirements, as have the non-parasitic adults of chiggers and also the aquatic larvae of mosquitoes. In such cases the pattern of these requirements is superimposed on the pattern of host distribution to make the distribution of the parasite even more patchy—but not to the degree one may expect because, in the course of evolution, parasites become adapted to their hosts in many ways including their favouring exactly the same habitat. Indeed, in many cases the parasite's adaptation to particular habitat conditions may be more restricted than its adaptation to feeding on various hosts. It is common to find a parasite in the field to be restricted in one locality almost exclusively to a particular host such as one species of rat. One is then surprised to find that the parasite in the laboratory feeds satisfactorily and completes its life-cycle on a great variety of other hosts which are also available in the neighbourhood, so that one may reasonably wonder why the parasite is not more widely distributed over a greater range of hosts and habitats. The reason is that the parasite is particularly adapted to a microhabitat shared by a particular host. In addition, of course, there may be something special about a particular host's behaviour that encourages or discourages a parasite. For example, in the case of a chigger that normally detaches itself after feeding for only a day or two, a flourishing mite-colony could hardly be built up by a host which ranged widely and did not return to the same place for intervals of days or weeks (see Fig. 4 and page 11). In Malaysia a population of field-rats living in grassland may be found immediately adjoining a population of house-rats living in huts. Each species of rat living in either different environment supports a very different pattern of parasites. In Singapore Island where the common field-rat has not yet become established, a population of house-rats may be found living in the grass, in which case they have the field-rat pattern of parasites. The implications in the case of man are obvious and of considerable concern to the epidemiologist. Many parasites are so dependent on the behaviour and habitat of potential hosts that they may be regarded as 'ecological labels', that is, their presence on a host conveys information about its ecology which may be very difficult to determine by the usual methods of trapping and observation in the field.[10]

Two more examples will illustrate this completely. Gan Island, Addu Atoll, in the Indian Ocean, is a mile long, with a swamp at one end and a village at the other. Its only land mammal is the introduced ship-rat, which swarms over the whole island. There are three species of trombiculids: the vector

Leptotrombidium deliense over most of the island except the swamp; *Ascoschoengastia indica*, a nest-dwelling species, in the village huts and to some extent in the coconut palm-tops; and *Blankaartia acuscutellaris*, confined to the swamp. The chiggers found on an individual rat will at once tell the observer where it has or has not been living for some days before capture. Of course, the trapper will already know this accurately. There will, however, be many occasions when novel information, impossible for the trapper to guess, may be inferred from the parasite-pattern. In Malaya, the tree-mouse, *Chiropodomys gliroides*, makes a hole in a giant bamboo, gnaws through the partition of an internode above, and makes therein a very clean little nest. It is never infested by chiggers. In Borneo, the corresponding species, *C. legatus*, was found by Traub to have two species of *Ascoschoengastia* which at once told us that it had a totally different kind of nest. We later found that it lived in tree-holes.

There are two ways of expressing this relationship between parasites and their hosts. One is that many or most parasites are so adapted to the microhabitat and behaviour of hosts that the pattern of parasitization of a population or even an individual will always be rich with information about the hosts. The other is that changes, often only slight, in the behaviour and microhabitat of a host will alter its parasite-pattern. The pattern of infectious and parasitic diseases is part of the total parasite-pattern. Add to this the fact that parasites may be indicators of stress in their hosts, and it is obvious that a study of parasite- and infectious disease-patterns *as wholes* is a most important tool in the hands of those concerned with preventive medicine or epidemiology. When the epidemiologist studying a particular infectious disease subdivides the population by sex and age-group, he is, among other things, simply dividing it up into groups that behave differently in relation to particular microhabitats and to other members of society. The ecology of each differing group is differently labelled by its parasite-patterns.

All complex problems should simultaneously be approached by as many different avenues as possible. The recognition of parasites as ecological labels gives us one novel biological approach to medical geography.[11]

Since the rabbit is the major host of *Neotrombicula autumnalis* in Britain, what would the situation have been there before the rabbit was introduced to Britain from Central Europe in the twelfth century? The answer is probably that the mite had already been introduced repeatedly by birds and that it was already being maintained in scattered patches by garden birds and various small mammals. The establishment of the rabbit gave the mites a new opportunity to be maintained at a higher level by a host whose size, behaviour, and habitat allowed dense colonies of mites to be built up, especially in the neighbourhood of warrens. Birds such as blackbirds are, however, still important hosts and have played a major role in recent exacerbations of infestation by the European harvest-mite in Denmark and Czechoslovakia, during which there have been many thousands of cases of severe scrub-itch.[9, 12, 13]

The situation in Britain has now gone a long way back towards the time, over 800 years ago, when no rabbits were there. The introduction of myxomatosis virus to rabbits in Britain in 1953 led to a huge reduction in the numbers of this pest, and this altered the character of the vegetation that had been rigorously controlled by the rabbits. It also led to repercussions among the animals that had preyed on the rabbits, and through these to other animals in the usual chain of events, which makes all types of interference with nature so complex, exciting, and sometimes frightening. One result has been an extensive reduction of infestation by harvest-mites where these were supported by rabbits. Birds and the other small mammals are however available to maintain the mites.

THE ACT OF FEEDING

Why should the harvest bug that Gilbert White writes about get into the skins 'especially those of women and children'? It is probable that women and children are more sensitive to the bites and therefore get more obvious and violent reactions. Thus, there may be many more complaints from women and children than from the men, even though the attack rate may be the same. It is possible that women and children may actually be more attractive to the chiggers. Beekeepers tell us that women tend to be stung more readily than men, and also that they tend to be stung more frequently when they are menstruating. Also, it may be that women and children are more prone to develop allergic sensitivity to chiggers.

After placing a number of unfed chiggers on their favourite host, it is not uncommon to find larvae wandering about for hours afterwards trying to make up their minds to settle down and feed. It could be that chiggers climbing on to a leathery-skinned man may take much longer to make up their minds to feed than if they had climbed on to a tender-skinned person, and the chigger that delays too long is likely to be killed at the time of the evening bath.

Not only are women and children likely to be more sensitive and possibly more attractive to the chigger bites, but they may be occupationally more liable to attack than the men at large. Children accompanied by their mothers may tend to play and picnic in infested sites, including gardens. The Pescadores Islands, which lie between Formosa (Taiwan) and the mainland of China, are low-lying islands of coral rock swept by strong sea breezes and spray. The inhabitants build protective walls of rock around their little gardens, in which are grown sundry vegetables and, of course, small patches of tall thatch-grass. The rocky walls provide an ideal habitat for rats, which forage in the garden and also frequent the shelter of the grassy patches. The rats encourage a species of chigger that happens to transmit scrub-typhus, and the chiggers are found in greatest concentration in the little grassy patches and around the walls. The men fish and the women farm. It is the women at work and the children at play who most often come into contact with these grassy patches and who have the highest incidence of infection.[14] This is a curious reversal of the usual situation in which the men working in the fields have the highest incidence of scrub-typhus.

Gilbert White twice mentions the effects of pest-chigger bites on man. The larvae do not burrow, but the swelling around them tends to bury them partially while they remain firmly attached by their tiny palpal claws. Chiggers have the curious habit of feeding through *stylostomes*, or drinking-straws, which form in the tissues as they feed, apparently as a product of their salivary secretions.[7, 15] These tubes may start forming within an hour or two of attachment and are left behind when the chigger detaches itself. Chiggers forcibly detached from animals after days of feeding may bring with them their stylostomes, sometimes as long as their own bodies, still attached to their mouthparts. The intolerably itching bumps raised in man are, however, due to allergic reactions to the saliva, the stylostomes apparently playing no particular part. Indeed, they seldom have time to develop far in man. Massive attacks by chiggers not only give rise to much suffering and loss of sleep, but they may be accompanied by a simultaneous non-specific fever to which White refers. More commonly, some degree of fever accompanies the secondary infection which follows scratching. These fevers are not to be confused with the fever of scrub-typhus which is a specific infection with an incubation period of about twelve days.

The pest-chiggers *Eutrombicula wichmanni*, *Schoengastia vandersandei*, and doubtless one or two other species were recognized as a potentially intolerable affliction by early explorers in New Guinea and the islands known as the Moluccas (see Appendix 2). *E. wichmanni*, widely known there as the *gonone* or *mokka*, occurs in greatest numbers in association with the large mound-nests of the bush-turkey *Megapodius*. Alfred Russell Wallace, who independently of Darwin conceived the theory of natural selection in the evolution of species, suffered severely from this pest: 'All the time I had been in Ceram' he wrote in 1869, 'I had suffered much from the irritating bites of an invisible acarus, which is worse than mosquitoes, ants and every other pest, because it is impossible to guard against them. This last journey in the forest left me covered from head to foot with inflamed lumps, which after my return to Amboyna produced a serious disease, confining me to the house for nearly two months.' This disease may have been due to secondary infection of the skin combined with exhaustion and loss of sleep. Wallace may, however, have contracted some coincident general infection, possibly even scrub-typhus.

Remarkably little is known of the development of sensitivity or immunity. There is evidence that sensitization leading to the formation of itching bumps or pustules may take as little as two weeks after first exposure, but many people show no response until towards the end of their first season of exposure, or even not until the following season. Although it is generally assumed that repeated exposure results in immunity there is no good evidence of this. Indeed, Poulsen found no evidence of acquired immunity in Denmark, although about a fifth of the population seemed to be immune without having suffered in the past. He found one person who became sensitized simultaneously to *N. autumnalis* and North American species of pest-chigger.[12]

THE OBSERVER VERSUS THE EXPERIMENTALIST

Like all good naturalists, Gilbert White was not only a keen observer but was able to pick out and present significant facts without wasting words. In 138 words he has given us information about the distribution and habitat of the harvest-mite, *Neotrombicula autumnalis*, its seasonal prevalence, its effects on man, the special liability of women and children to develop the rash, and the fact that infestation appears to be an occupational hazard.

The fact that these mites attack man would seem to suggest that they must normally be parasitic on some animal, but this inference need not necessarily be correct because we now know that rashes due to a quasiparasitic attack may be produced on man by various other mites, even ones that normally feed on plants or pollen (see Appendix 3). Our first assumption that the harvest-mite may be an external parasite prepared to attack man can easily be tested. The mites occur in such large numbers in and around rabbit warrens that they discolour the nets of warreners and, as an observer has put it, add a reddish cast to the barrel of a gun laid on the ground. If these mites are parasitic, then the rabbit must surely be one of their favourite hosts. It would be very easy to take a clean rabbit, place it on the ground and observe mites crawling onto it, later to find that the mites have attached to the ears and body of the rabbit in large numbers and that the appearance of the rabbit is exactly the same as that of a rabbit found naturally infested in the warren. By keeping the rabbit under observation it would easily be found that within a week most of the attached mites which started as 6 scampering legs carrying a tiny body like a speck of red dust had become replete little balloons which detached themselves, stumbled towards the ends of hairs, and fell to the ground where they sought shelter. There each mite would stretch out its legs and become a pupa, and within a week or two a totally different kind of velvet-mite with 8 legs would emerge. The enquiring observer could then easily have gone to the infested garden and trapped voles, blackbirds, and thrushes, as well as rabbits, and would have found other natural hosts for these parasitic mites. He could also have found 8-legged nymphs and adults in the soil.

Gilbert White did none of these very simple things. He did not even comment on the infestation of rabbits in the field. From his letters, it can be seen that this was due to the nature of his particular curiosity. Gilbert White was a well-educated gentleman, happy in his beloved countryside. He was a great lover of nature, of the flowers, trees, birds, and bees; and a very keen observer. Insects were certainly among the creatures that interested him and he even carried out a trivial experiment in transporting crickets. Reading through *The Natural History of Selborne* one is struck by the vast amount of observation about nature, especially birds, and by the many thoughts these observations set in motion; but there is no evidence of probing nature to get answers. Contrast Gilbert White with someone like that great naturalist, Henri Fabre, or with White's own mentor, the inventive physiologist, Stephen Hales.[16] These latter gentlemen would surely have done just what we have described and would have brought both harvest-mites and animal hosts into their studies for observation and experiment.

What makes the difference between people such as Gilbert White and Henri Fabre or Stephen Hales? All were intensely curious people but White was satisfied with observation in what some would regard as an idle fashion. Could it have been due to differences in their education? And what did their education comprise? Hale's friend, the Rev William Stukeley, M.D., was a few years behind Hales at Cambridge University and has recorded some of his college education thus:[17]

I was matriculated Spring 1704. I staid all that year in College, applying myself to the accustomd studys, & constantly attending Lectures, sometimes twice or thrice a day, & Chappel thrice a day, & scarce missed three times all the while I staid in College. My Tutor, & Mr., now Dr. Danny, . . . joined in reading to their respective Pupils. The former read to us in Classics, Ethics, Logic, Metaphysics, Divinity, & the other in Arithmetic, Algebra, Geometry, Philosophy, Astronomy, Trigonometry. Mr. Fawcett read to us in Tullys offices, the Greek Testament, Maximus Tyrius by Davis, Clerks Logics, Metaphysics, Grotius de jure Belli & Pacis, Pufendorf de Officio Nominis & Civis, Wilkins Natural Religion, Lock of human Understanding, Tullys Orations. Mr. Danny read to us in Wells Arithmetica numerosa & speciosa, Paradies Geometry, Tacquets Geometry by Whiston, Harris's use of the Globes, Rohaults Physics by Clark. He read to us Clarks 2 Volumes of Sermons at Boyles Lectures, Varenius Geography put out by Sr Isaac Newton, & many other occasional pieces of Philosophy, & the Sciences subservient thereto. These courses we went thro with so much constancy that with moderate application we could scarce fail of acquiring a good knowledge therein.

Gilbert White would have had a similar exposure at Oxford where, however, the tradition was classical and less scientific. Clarke[18] flatly says that, in the eighteenth century, 'At Oxford the degree exercises had sunk to a mere farce, so that tutors were at liberty to devise their own lecture courses, or leave their pupils to themselves.' (A tutor was evidently too lowly a person to have responsibility for what he taught: he passed on the words of the masters.) By the time White went to Oxford, it was customary to write a theme, declaration, or translation every week. In those days there was one broad type of education for a gentleman, on which was superimposed some special professional training for those who wished to follow religion, medicine, or the law.

One may perhaps liken this to being introduced to a river of knowledge which flows like wine, wine which one may taste to satiation. It is there to be tapped at any moment. It is as if one were to lie back and quaff at leisure, so that knowledge may seep in and by thinking be made to permeate one's philosophy. There seems to be something passive about this kind of education. It is almost as if it were something done to one, like vaccination. Furthermore, being surrounded by so much information capable of assimilation, there must be a natural tendency to feel that being knowledgeable is alone the test of education. This kind of handling of knowledge is nowadays known as data-processing, ably carried out by computers. The facile acceptance of reported information is no longer fashionable. It is indeed so unacceptable that we are astounded, for example, at the gullibility of many famous philosophers and travellers in the past. (Aristotle stated as a simple fact that women have fewer teeth than men, obviously without putting this to the test, not even by looking into the mouth of either of his two wives.) Unacceptable though unsupported facts may be, the massive absorption of 'acceptable facts' is now very popular, at the expense of mental indigestion.

Oxford nevertheless was the birthplace of the experimental method, at least in Britain, thanks to the astonishing contributions of Robert Grosseteste, followed by Roger Bacon.[19] It seems that their teaching inspired only a few naturally receptive people in White's time, five centuries later.

In the eighteenth century, the fundamentally significant parts of knowledge were limited and it was possible for an intelligent person with opportunities to comprehend most of what mattered about most known things. Now the river of knowledge has not only swollen to a flood, but it has spread over a vast delta so that one seems for ever to be stranded on some islet trying to get across to another piece of solid ground. In these circumstances, data-processing by the student is seen by many to be the essence of education. The student responds by being happy when he receives what can be written down in his notebook, but restless when called upon by his professor to slow down his pace and to digest and think about principles. There is a tendency for the science student to carry this data-processing to excess simply because there is so very much 'essential' information that seems to demand assimilation. The contrary tendency is found among some students of humanities who are prepared to assimilate a great deal of material which need not be factual at all, and with this to speculate philosophically without the strict discipline of constantly checking hypothesis against reality. Among them, the realities are values and judgements about them are subjective.

Although medical students are drawn from the humanities as well as from the sciences, their medical training is scientifically based. It is noticeable among them that many who are scientifically sound, responding well to the data-processing of the first two years, become relatively lost when called upon to respond sensitively to and to communicate with patients in the third and subsequent years. (The opposite also happens, with an unexpected improvement in the third year.) They do not realize that the satisfaction they gain from investigation, making a diagnosis, and prescribing treatment, is the satisfaction of having processed data computer-wise. The more they have absorbed the scientific spirit, the more do they tend to concentrate on the patient's physical and chemical aspect, which is so much more readily manipulated than the spiritual aspect of the patient as a person. This is among the processes that together tend nowadays to produce the medical scientist rather than the healer. One result in the field of medical education is to narrow the focus on medicine so that attention is distracted from the more subtle and complex psychological and ecological processes, towards the more comprehensible and readily applicable molecular aspects. This is turn brings the weight of interest onto treatment, and to a lesser extent protection, of the individual. Interest tends to be taken off collective medicine and human ecology. Finally, within the field of preventive medicine, attention tends to be focused on the scientifically based aspects of environmental sanitation and immunization procedures, to the neglect of the much more subtle but vitally important sociocultural aspects. These tendencies have been changing rapidly in the last decade or so. The modern sharpening of interest in ecology, the broadening of ecological research, and the introduction to it of quantitative methods, are indications that the basic sciences have progressed far enough for scientists to reach towards the complexities of social systems and ecosystems. Ecology may be described as the physiology of such systems, studied as wholes. I shall return to this in my closing remarks.

Gilbert White's attitude towards the world around him was doubtless a product of a passive education on a passive scholarly personality. His attitude, unchanged perhaps because of lack of example among his teachers (but not among his friends), did not lead him to probe for hidden knowledge, the wine in sealed bottles. White may be contrasted with his friend Stephen Hales but he cannot be accused of lacking curiosity. It was lively curiosity that led White to gather such a vast amount of personal observation and local lore. But, it was in the operation of that curiosity that the two differed. Hales had found that there was something better than quaffing from the river of wine, and that was to take trouble to find and open the sealed bottles. One was passively observing and soaking in knowledge; the other was actively probing and evoking it. This is largely the difference between the philosopher and the experimentalist, the latter more restlessly critical than the former—I don't mean critical in logic but critical in accepting the face value of appearances. The difference is analogous to that found between the humanist and the scientist.

In referring to 'the scientist', we must not confuse him with the technician. Good technicians can be good scientists, and vice versa, but there is a great difference between the work and outlook of the technician, concerned in the execution of scientific business, and the true scientist, who is a philosopher as well. Also, it is true that science offers some people an escape from reality so that some scientists are complete cranks. Finally, the narrow scientist—and there are many of them—however brilliant he may be in his special field, is not necessarily to be trusted outside it.

THE ECOLOGICAL OUTLOOK

Since these differences involving science and the humanities are already presenting us with the greatest educational dilemma in the history of man, it behoves us to study them from as many angles as possible. The angle from which an individual views life and its problems is popularly known as his 'slant'. In spite of their basic scientific disciplining, scientists may have very different 'slants'. I am convinced that there is a distinct ecological 'slant' by which some people will see a given situation in a distinctive way, difficult to comprehend by many others. On the other hand, these others may be seeing the situation in ways to which the ecologist is partially blind. I once told my friend J. L. Harrison, who is mathematically minded, that I had advertised my car for sale, at the 'nearest offer to £380'. He instantly countered with the point that if I were offered £370 by one and £400 by another, I must take the £370. Of course, everyone knows that such advertisements are conventionally worded and do not mean exactly what they say. But only someone with what we might call a precise 'statistical slant' would instantly think of this interpretation.[20] An excellent way of discovering scientists with differing 'slants' groping towards each other's viewpoints or *Anschauungsweisen* is to read the verbatim discussions at wide-ranging conferences, such as the Josiah Macy conferences on cybernetics.[21]

What, then, is this 'ecological slant'? And indeed, what is ecology, about which we hear so much but seem to know so little? There have been some recent reviews of this subject, all agreeing that the essence of ecology is the study of entire systems as wholes.[22, 23] A *system* in this definition is an assemblage of interacting and interdependent component parts, comprising living and non-living components and all media manipulated by them or through which information is exhanged between them. A system at this level is known as an *ecosystem*, to distinguish it from such a system as an individual organism. We may therefore conceive of a series of studies, each concerned with a higher level of organization than the one before:

1. *Physiology*, at the level of integration of organs and tissues comprising an individual organism.
2. *Autecology*, at the level of the organism itself (or of a species), in relation to its environment. Autecology (*auto*, pertaining to oneself), coined by Schröter in 1896, is distinguished from *synecology* (*syn*, together) or the ecology of assemblages of mixed species. Adams in 1913 used 'individual ecology' and 'associational ecology' for these somewhat cumbersome terms. I am unable to find any difference in meaning between autecology and *bionomics*. It is only in autecological or bionomic studies that the term 'environment' has meaning. An example is the autecology of the Norway rat.
3. *Population (syn)ecology*, at the level of the species-population or deme, e.g., the population dynamics of the vole or lemming.

4. *Community (syn)ecology* (or *biocenology*), at the level of what is known as a species-network of interrelated food-chains, one organism feeding on another or its products.

5. *Ecosystem (syn)ecology*, at the level of entire ecosystems comprising both living and dead matter. In a way, this is the physiology of ecosystems. 'Ecology' unqualified should refer only to this.

Our thinking about ecology would probably be improved if we reserved the term ecology for study at the level of the ecosystem, and used the term autecology or bionomics for the necessary but fragmentary studies of the 'ecology' of the Norway rat, the lemming, or even man (human (aut)ecology). In the case of man, however, the complexities of his social system simply add a new dimension to the ecosystem of which he is a part. Man is a dominant animal weed capable of shaping and even disrupting the ecosystem—the significance of animal weeds is discussed later, in Chapter 4. Human ecology may, therefore, be studied synecologically.

If we speak of man-made smog as something 'in our environment', we are not thinking ecologically. If we think of smog as a metabolite in fluids of the metropolitan ecosystem, analogous to pyruvate or lactate in the fluids of the body, then we are thinking ecologically. There is inherent in all fragmentary ecological studies a serious hazard of failing to see the wood for the trees. Of course we must have many autecological studies, but never so as to allow the ecosystem to be forgotten. Another analogy helps us to understand this: a physiologist who studies the seawater in our veins or the breezes in our bronchi has the entire body at the back of his mind, but those pursuing autecological studies frequently ignore the ecosystem of which they are studying a fragment. One reason is that the 'entire body' of an ecosystem is a frightening complex entity of which we are mere components; another is the difficulty of melding living with nonliving components; and another is that many people do not like the idea of being part of some sort of superorganism with a life of its own, for this tends to strike at our ideas of human freedom. Yet another reason is that we can get very good practical results by autecological research, even though we can get ourselves into appalling difficulties when we try to be clever on a large scale. The man-made maladies which Rachel Carson brought brutally to the fore are the side-effects of applying autecological studies, with such success that millions upon millions of mouths are well-fed or even overstuffed. Much of the debit side of this balancesheet is due to application of such research without enough studies at the level of the ecosystems.

'Environment' is a term belonging to fragmentary ecological studies, but inapplicable to studies of ecosystems in which a house or nest, a rat and its parasites, an automobile and its exhaust, the air and the water and the plants and the soil are all components in a system. 'Environment' is a concept which appears only when one component is artifically isolated for study—temporarily, we would hope.[22]

In preventive medicine or public health, concerned with an important fragment of man's self-created problems, it is necessary to get together research workers with as many different 'slants' as possible. Among these, the ecologist's outlook is not merely important but essential.

REFERENCES

1 ZINSSER, H. (1934). *Rats, Lice, and History. Being a Study in Biography, which, after Twelve Preliminary Chapters Indispensable for the Preparation of the Lay Reader, Deals with the Life History of Typhus Fever.* Boston, Little Brown. [Frequently reprinted. A recent paperback edition is in the Bantam Classic Series, Bantam Books, New York, 1960.]
2 AUDY, J. R. (1960). Evolutionary aspects of trombiculid mite parasitization, pp. 102–8 *in* R. D. PURCHON, Editor, *Proc. Centenary & Bicentenary Congr. Biol., Singapore, 1958.* Singapore, Univ. of Malaya Press, and Cambridge University Press.
3 HESSE, P. von (1899). Die Ausbreitung des Sandflohs in Afrika. *Geogr. Z.* **5**, 522–30.
4 FORD, J. and HALL, R. de Z. (1947). The history of Karagwe (Bukoba District). *Tanganyika Notes* **24**, 1–27 [Includes explorer's accounts of human devastation during the first advance of the chigoe flea].
5 HOEPPLI, R. (1963). Early references to the occurrence of *Tunga penetrans* in tropical Africa. *Acta Tropica* **20**, 143–53.
6 AUDY, J. R. (1951). Trombiculid mites and scrub-itch. *Australian J. Sci.* **14**, 94–6. (See also *Trans. roy. Soc. trop. Med. Hyg.* (1952), **46**, 459–60.)
7 HOEPPLI, R. and SCHUMACHER, H. H. (1962). Histological reaction to trombiculid mites, with special reference to 'natural' and 'unnatural' hosts. *Z. Tropenmed. Parasitol.* **13**, 419–28. [Which adduces no support for the hypothesis but opens up the inquiry into histopathology of trombidiosis in animals.]

8   EWING, H. (1923). Our only common North American chigger, its distribution and nomenclature. *J. agric. Res.* **26**, 401–3.
9   TUXEN, S. L. (1950). The harvest mite, *Leptus autumnalis*, in Denmark. Observations made in 1949. *Entomol. Medd.* **25**, 366–83.
10 AUDY, J. R. (1960). Parasites as 'ecological labels' in vertebrate ecology, pp. 123–7, *in* R. D. PURCHON, Editor, *Proc. Centenary & Bicentenary Congr. Biol., Singapore, 1958.* Univ. of Malaya Press, and Cambridge University Press.
11 AUDY, J. R. (1954). A biological approach to medical geography. *Brit. med. J.* i, 960–2. Also: (1958) Medical ecology in relation to geography. *Brit. J. clin. Practice* **12**, 102–10.
12 POULSEN, P. A. (1957). *Undergelser over* Trombicula autumnalis *Shaw og* Trombidiosis i Danmark. *Studies on* Trombicula autumnalis *Shaw and* Trombidiosis in Denmark. *With English Summary*. Denmark, Universitetsforlaget i Aarhus, 149 pp. [Intensive study of Thisted, following Tuxen.]
13 DANIEL, M. (1961). The bionomics and developmental cycle of some chiggers (*Acariformes, Trombiculidae*) in the Slovak Carpathians. *Czechoslovenska Parasitol.* **8**, 31–118. [A very thorough study.]
14 COOPER, W. C., LIEN, J. C., HSU, S. H. and CHEN, W. F. (1964). Scrub typhus in the Pescadores Islands: an epidemiologic and clinical study. *Amer. J. trop. Med. Hyg.* **13**, 833–8.
15 SCHUMACHER, H. H. and HOEPPLI, R. (1963). Histochemical reactions to trombiculid mites, with special reference to the structure and function of the 'stylostome'. *Czeskoslovenska Parasitol.* **14**, 192–208.
16 CLARK-KENNEDY, A. E. (1929). *Stephen Hales, D.D., F.R.S. An Eighteenth Century Biography*. Cambridge University Press.
17 CLARK-KENNEDY, A. E. *Ibid.*, p. 7, quoting W. C. LUKIS, Editor (1882), *Family Memoirs of the Rev. William Stukeley, M.D.*, 3 vols., London, Surtees Society.
18 CLARKE, M. (1959). *Classical Education in Britain, 1500–1900*, pp. 67–73. Cambridge University Press. See also C. WORDSWORTH (1874). *Social Life at the English Universities in the Eighteenth Century*. Cambridge, Deighton, Bell & Co. B. SIMON (1960) *Studies in the History of Education 1780–1870*. London, Lawrence & Wishart.
19 CROMBIE, A. C. (1953). *Robert Grosseteste and the Origins of Experimental Science, 1100–1700*. Oxford, Clarendon Press.
20 AUDY, J. R. (1958). The localization of disease with special reference to the zoonoses. Discussion. *Trans. roy. Soc. trop. Med. Hyg.* **52**, 308–34 (pp. 333–4).
21 FOERSTER, H. von, Editor (1949–1952). *Trans. Sixth* to *Trans. Ninth Conference on Cybernetics*. New York, Josiah Macy, Jr. Foundation.
22 AUDY, J. R. (1965). Human ecology *in* Emerging disciplines in the health sciences and their impact on health science libraries. *Bull. med. Lib. Ass.* **53**, 410–19.
23 SARGENT, II, F. (1965). Guest Editor, papers on Human Ecology. *BioScience* **15**, 512–26.

2

AKAMUSHI:
THE RED MITES OF JAPAN

THE northwest part of the island of Honshu, Japan, supports a peculiar type of mixed deciduous forest[1] which indicates a very distinctive combination of climate and terrain. Within this one may find the northernmost outpost of an arthropod that has become the most notorious in Japan: the *aka-mushi*, red-insect or red-mite.[2] It is a chigger, a trombiculid mite, with the scientific name of *Leptotrombidium akamushi*,[3] very closely related to an even more widespread species, *L. deliense*. Farther south, down to Australasia and the Malay Archipelago, the *akamushi* occurs widely but in Japan it is restricted to particularly favourable sites for reasons similar to those described by Ewing for the North American pest-chigger *Eutrombicula alfreddugesi* in the northernmost part of its range (Chapter 1, page 11). Many creatures and plants at the limits of their geographic range are similarly distributed in scattered 'islands'. The *akamushi* in Japan is now confined to a few endemic areas in the flood-plains of three rivers in the prefectures of Niigata, Yamagata, and Akita. It has other local names such as *tsutsuga-mushi*, 'dangerous-mite', so named because it transmits a serious rickettsial infection; *shima-mushi*, or 'island-mite', so named because it frequents the grassy islets and banks of fertile silt which are deposited where the rivers debouch onto the plain; and *ke-dani* or 'hairy-mite' (tick), which might almost be the equivalent of our 'velvet mite'.[4]

This red-mite, the akamushi or tsutsugamushi, is not a pest-chigger, but it does transmit the rickettsia of scrub-typhus, a

[1] Found again in only one other place: Maine, USA.

[2] See Appendix 4 for an account of the Japanese names for chiggers and scrub-typhus.

[3] Until recently known as *Trombicula akamushi*—see Appendix 1.

[4] See Appendix 5 for major publications on scrub-typhus and epidemiological studies of the vectors; also Appendix 3 for references to chiggers and Appendix 4 for Japanese names.

form of typhus also known as tsutsugamushi disease, Japanese river fever, or flood fever. This infection, causing a high mortality among Japanese farmers, has been studied with extraordinary intensity by Japanese research workers for nearly a century. Tsutsugamushi disease has become part of the lore of Japan in much the same way as the similar typhus fever, Rocky Mountain spotted fever, has become part of the lore of the western USA. One of the Japanese pioneers in research, Norio Ogata, wrote the following in an historical account of tsutsugamushi research in the *Tokyo Medical Journal* in 1953:[1]
... some of the older workers in this field—Nagayo, Kawamura, and Tanaka—died several years back, and Hayashi died last May. I am now the oldest one left and feel lonely in my old age. ... In the footsteps of the older workers, I started studies in this field in 1916 and have been studying tsutsugamushi disease ever since. I have lived in this marsh for thirty-eight years. Moreover, my father, Masanori, was engaged in the study of this disease for sixteen years, from 1904 to 1919; my nephew, Manabu Sasa, is presently looking for new species of tsutsugamushi; and my oldest son, Takayuki, is now working in the Rickettsial Section of the Institute of Infectious Diseases in Tokyo University. Therefore, our family has been engaged in the same study for three generations. No one has ever asked his son to go into the field, but strange as it may seem, it happened. Our family seems to be doomed to have some special affinity for tsutsugamushi disease.[1]

EARLIEST ACCOUNTS

This disease appears to have been more widespread in the past, but to have disappeared from several areas where it was formerly present, presumably because of changes in land-usage and river topography and, more recently, flood-control measures. For example, Hatori[2] wrote that in the year 1840

[1] I am grateful to Miss Naila Minai for translation of this paper, to Dr Nobuo Kumada for his advice on difficult passages and discussion of background, and especially to Dr Cornelius B. Philip for making available to me a translation No. 4-6-63 by the Translating Section, Library Branch, Division of Research Services, National Institutes of Health, Bethesda, Maryland, USA. Dr Norio Ogata has also kindly answered a number of questions in correspondence. His review is a valuable document. I am grateful to Dr Philip for permission to quote from the translation. Drs Nobuo Kumada and Manabu Sasa have also helped me to understand nuances.

one Genkei Ohtomo, physician to the lords or Daimyo, had a reputation for curing tsutsugamushi disease. This was in the Yonezawa district in a place near Akita where this disease was unknown later in the century. Also, shrines put up to placate the vector mite and appeal to it to spare the farmers (Figs. 8, 9) have been found in places where the infection no longer troubles people and is even difficult to find among the small rodents.

Many centuries ago a 'Pine Festival' was instituted at Haguro, Yamagata prefecture, at which effigies of the tsutsugamushi mite have been burnt for at least two centuries. At that latter time, during the feudal era, many farmers were dying of the disease which was correctly ascribed to the small vector

FIG. 5. A *tsutsuga* effigy (of an adult vector mite) at a New Year *Mushi-yaki* (mite burning) ceremony at Haguro (see also Appendix 7). Figs. 5–9 were drawn by Miss Marjorie Smith from black-and-white or colour slides, several photographed by Dr Kiyoshi Asanuma, given to the author by the late Dr Takeo Tamiya and by Dr Masami Kitaoka. Reproduced with their permission.

in the fields. Prayers were regularly offered at shrines. The story told to me by the late Takeo Tamiya is that on the tenth day of prayer by a particular farmer who was seeking guidance, God appeared and instructed that the fields should be burnt. (We now know that burning has no lasting effect on the vectors, although, by encouraging drying-out, it temporarily reduces the risk of infection.) From that time, hay and ropes have been made into two very large figure-of-eight bundles representing the adult mites and called 'tsutsuga' (Fig. 5). Every January at the Haguro Shrine these two effigies are burnt at the festival—the *mushi-yaki* or mite-burning—as a sort of competition to see which makes the better bonfire. The binding ropes are removed beforehand and used by the villagers to make their own tsutsuga effigies, to be placed under the eaves above the doorways of their houses as good-fortune charms (Figs. 6, 7). The Haguro Town Office has one of these

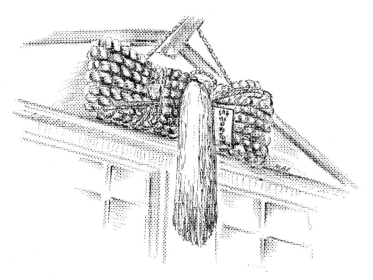

FIG. 6. *Tsutsuga* effigy under eaves of a house near Haguro, with name of the family, made of rope from the *Mushi-yaki* ceremony. From fig. 2, p. 16, T. Tamiya, Ed. (1962) *Recent Advances in Studies of Tsutsugamushi Disease*. Medical Culture Inc., Tokyo.

Fig. 7. A small *tsutsuga* effigy under the eaves of a house, Haguro. Photo by K. Asanuma.

over its front door, in a glass case. This western part of Yamagata Prefecture has been free of the tsutsugamushi disease (scrub-typhus) for a very long time, and what used to be an annual rite to protect people from the dangerous mite is now a cheerful festival to bring good fortune and ward off fire and sickness.

A fine monograph on recent studies of tsutsugamushi disease was published in 1962 under the editorship of the late Takeo Tamiya.[3] After describing the annual festival at Haguro, the authors draw attention to various shrines which have been put up to ward off the vector mite (the italic numbers in square brackets within the quotation refer to figures in the present Lectures[1]):

Another example is Kedani-Daimyojin Shrine in Arato, Yamagata Prefecture, along the upper course of Mogami River (Fig. 4[8]). For at least 50 or 60 years no case of Tsutsugamushi disease has been found there, but the shrine was established in the year 1860, according to a plaque inside. The name Kedani-Daimyojin means 'hairy mites are enshrined' so as not to cause harm to human beings. So it is certain that this area was formerly a very noxious area where many farmers died after being bitten by the mite. On the back of the plaque there is an inscription: 'Let us get rid of all noxious mites' (Fig. 5).

[1] I am very grateful to the late Dr Takeo Tamiya and to Dr Masami Kitaoka for their kindness in giving me colour slides and photographs, from some of which Figs. 5–9 have been drawn. Some of these photographs were taken by Dr Asanuma.

AKAMUSHI: THE RED MITES OF JAPAN

Here and there in Yamagata and Niigata Prefectures we see stone monuments inscribed with the words, Akamushi-Daimyojin, meaning 'red mites are enshrined here' (Fig. 6).[1]

FIG. 8. Kedani-Daimyojin (hairy-mites-enshrined) Shrine in Arato, Yamagata Prefecture. Established 1860, but no cases of scrub-typhus have been found for at least sixty years in this upper region of the Mogami River. The shrine contains a plaque with the inscription 'Let us get rid of noxious mites'. Same shrine as fig. 4, p. 17 of Tamiya (1962) but from a different angle. From photo, K. Asanuma.

All of these examples show the transitory nature of Tsutsugamushi disease, with formerly endemic areas becoming safe, and formerly safe areas becoming dangerous.

Fig. 9 shows another shrine, at the junction of the Mogami and Sagae Rivers in Yamagata Prefecture.

[1] There are, of course, subtleties in the rendering of terms such as 'Akamushi-Daimyojin' into English, so that an explanation is often necessary even of simple translations.

Fig. 9. Akamushi-Daimyojin, shrine at the junction of Mogami and Sagae Rivers, Yamagata Prefecture. From photo, K. Asanuma.

Ogata, *loc. cit.*,[1] refers to a shrine in Akita associated with a different kind of ceremony:

The community of Shimokadoma, which is a part of the village of Benten in Okatsu-gun in Akita prefecture at the junction of the Minase and Omono Rivers, has long been known as a habitat for tsutsugamushi and has suffered from them since olden days, producing about 500 deaths in this village alone. [Over what period of time is not stated.] On the outskirts of the village is a *Kedani Jizo*[1] temple, the only one of its kind in Japan. Every year on the 25th and 26th of July the villagers gather in front of the temple in

---

[1] *Jizo* is Bodhisattva, among other things the patron saint and guardian of the souls of dead children, and the patron saint of travellers. This particular shrine differs from those described by Takeo Tamiya and his colleagues by being essentially a *Jizo* shrine.

remembrance of those who died of the disease. The crowds come to watch the *Fuda-nagashi* [a ceremony of throwing symbolic tickets into the rivers].

The temple has a long history: about 200 years ago, a Buddhist priest, Tetsumon Shonin of Dewa, came to the place and was concerned by the great suffering due to tsutsugamushi disease. He made a thousand tickets which were believed to have the power to inhibit the activity of the tsutsugamushi mite, gave the tickets to the villagers and asked them to throw three tickets into the river every year. He told them that this would lessen the activity of the mites and that when the thousand tickets were exhausted the mites would be exterminated. The villagers have faithfully followed his instructions. The principal Buddhist statue in the temple is a carved wooden *Jizo*.

It must be admitted that this priest was cautious. It speaks well for the patience and discipline of the people who followed his advice with the prospect of relief to come after three centuries.

I am grateful to my friend Dr Nobuo Kumada, who recently spent 18 months with me in the G. W. Hooper Foundation, for a copy of a hitherto overlooked account of tsutsugamushi disease by a sanitary engineer, Tokibumi Kakurai.[4] It was published 'not for sale' in 1915 as a public health educational gesture by the Police Department of Yamagata Prefecture (see Appendix 7). A *kedani-myojin* is reported at Kono village on the banks of the Mogami River. Another is described in Arashi town, with full address, about 7 ft. high with a thatched roof and surrounded by a stone wall; and with the comment that 'to this day there is a special office that cares for the historic *kedani-daimyojin* [*dai*, great]. Every year on March 15, five most prominent benefactors make a pilgrimage to the shrine to perform sacred rites, placing wooden placards on the shrine. These placards include respects to the *kedani-daimyojin* and prayers from the poor mortals for peace, tranquillity, and justice in their homes and the nation.'

This same account refers to a local belief that the tsutsugamushi was a *dani* of a great serpent living there. Also near Houra village there is a shrine for a venomous serpent-god, in honour of a great serpent that lived there in olden days.

Apparently, some time before the mid-nineteenth century there was a belief that the noxious *mushi* was born out of the scales of venomous serpents. The dangerous viper *Agkistrodon* is incidentally known as *mamushi*.

From these accounts we learn that perhaps four or more centuries ago, an arthropod was recognized as causing a systemic disease, and that for at least two centuries Japanese farmers and medicine-men have been using the same name, *tsutsugamushi*, 恙虫, 'noxious bug', (or its earlier dialect form *tsutsuga no mushi*, the insect-of-danger), for both the vector mite and the disease (*tsutsugamushi-byo*). *Tsutsuga* offers little difficulty in translation. In writing it is represented by two ideographs which are pronounced the same but have slightly different meanings. 恙 alone is *tsutsuga* or *yō*, meaning illness, harm, or danger. In this sense, *tsutsuga* forms part of a formal farewell, especially for someone going on a journey: *tsutsuganaku* or 'safely', which means '(be) without danger' and could be loosely translated as 'fare thee well' or 'safe journey!' Those who suffer committee meetings will be pleased to know that *tsutsuga* may also be used to state that a meeting has adjourned without mishap. The other ideograph, for *tsutsuga*, 獇, is the same but with the addition of a radical signifying 'beast'. This *tsutsuga* is the ferocious lion or lion-dog that is supposed to eat men, tigers, and leopards. It is the *chinthe* of Burma, after which the Chindwin River and the guerrilla Chindits are named. Visitors to the Nikko Shrine will pass through the Yomei gate to find another called Karamon, on the top of which is a pair of animals facing each other: these are *tsutsuga* or 獇.

Mushi may be translated as insect, but also worm, bug, or—why not?—mite. The ideograph is supposed to have been derived as follows from an indubitably worm-like creature:

It is difficult to trace the recognition of the disease and the tsutsugamushi vector back with certainty for more than a few centuries in Japan. In China, however, there is good evidence of established beliefs as far back as the sixteenth century. Louis

Sambon in 1928 gave a delightful account of how he followed up a reference by Juro Hatori[2] to the 'Sand-mite':[5]

Now Hatori quotes a passage from a Chinese work of the sixteenth century, entitled 'Honzo Komoku' (System of Natural History), edited by Li Shiting,[1] in which a 'Sand-mite' is described as a fever carrier.

This historical epidemiological statement was so surprising that I called on Dr. Giles, Deputy-keeper of the Oriental Department in the British Museum, and asked him whether the work mentioned could be found in Europe. He told me it was in the room backing the one in which we sat, if, as he supposed, my title was the Japanese rendering of the name of a well-known Chinese 'Materia Medica', the *Pên ts'ao Kang Mu*, compiled by Li Shih-ch^{e}n ... [who] spent thirty years on the work, making abstracts from the writings of no less than eight hundred previous authors.

I asked to see the work. 'Come along', said Dr. Giles, 'but how are we going to find your mite; you have no page reference and the work consists of fifty-two volumes!' I looked in wonderment at the imposing array, drew out a volume and opened it. The ornate Chinese script was arranged in two columns on each page made of felted mulberry-bark fibres. 'And you can read and understand this?' I asked, placing my forefinger upon the fourth column. He began to read, then, suddenly turning, he asked for my paper, because he had chanced upon 'Sand-mites'—we had found the very passage I needed—and he gave me the following translation:

'Li Shih-chên says: "According to the Kuang chih of Kuo I-hung, the *sha-shi* (lit. 'Sand-louse') lives on the water. Its colour is red, and it is not larger than an ant. It kills man by burrowing into his skin."

'Ko Hung says, in his *Pao p'o tzŭ*: "The shih is found both on water and on dry land. When people walk on the sands after rain in the early morning or evening, it attaches itself to some part, such as their hair, and piercing the skin, penetrates into the body. It can be pricked out with a needle. In colour it is red, like cinnabar. If it is not extracted, it will enter the flesh and may cause death. In all places where this insect is encountered, the cautery should be resorted to on one's return home, and the heat will expel it."

[1] *Li Shiting*, more properly *Li Shih-chên*, is *Rijichin* in Japanese. The Chinese ideographs for *Pên ts'ao Kang Mu* are pronounced *Honzo Komoku* in Japanese. It can readily be seen that experienced guidance is required in tracing references, and that the original ideographs should be reproduced in all serious works. The work referred to is best described as a herbal.

'The *Chou hou fang* (a book of medical prescriptions) says: "Among hills and water, 'Sand-mites' abound. They are so minute as to be hardly visible. When people wade through water, or walk amidst undergrowth in the dark, these mites fasten upon them and burrow under the skin, causing prickly sensations and a red rash, the size of millet grains. Three days after the piercing of the skin, fever and ulcers supervene, and if the insect makes its way into the bones, death will ensue.

' "The people of Ling-nan (a name for Kwangtung), when attacked by the insect, scrape it out with the blade of a rush or a leaf of bamboo, and then smear the place with the juice of k'u-chü (*Lactuca* sp., or *Cichorium endivia*). But if it has already gone deep, it is extracted with a needle. The insect is just like a chieh-ch'ung (lit. 'itch-mite')." '

Li Shih-chên goes on to describe the symptoms of the disease—headache, rise of temperature, vomiting, numbness in fingers and toes, sometimes abdominal pains, mental depression, etc., which leave little doubt as to the nature of the malady.

This last point is particularly important because it is almost certain that there was some confusion between the true typhus infection by rickettsiae from species related to *Leptotrombidium akamushi*, and scrub-itch with or without non-specific fevers caused by heavy attacks of pest-chiggers—fevers such as those mentioned by Gilbert White, men tending their nets over rabbit-warrens being 'so bitten as to be thrown into fevers'. Sambon, *loc. cit.*,[5] pp. 101–3, perpetuates a similar error in his discussion of the Aztec terms Matlalsahuatl and Tlalsahuatl which have been so used as to allow confusion between epidemic louse-borne typhus, endemic flea-born typhus, and attacks by the pest-chigger *Eutorombicula alfreddugesi* in Mexico (see discussion of terms in Appendix 4).

THE TSUTSUGAMUSHI[1]

The names *tsutsugamushi*, *akamushi*, and *shimamushi* (with *tsutsugamushi-byo* for the fever) are common in the Niigata Prefecture, while *kedani* and *shashitsu* (sand-mite, from the Chinese) with *shashitsu-byo* for the fever, are preferred in the Akita Prefecture. It would be a mistake to suppose that these

---

[1] See Appendix 4 for Japanese and other names for chiggers and scrub-typhus.

names were regularly applied to specific identifiable mites. Indeed, in many cases a belief in the existence of these mites existed among those who had never seen them. Confusion was actually increased when the mites first came to be studied more scientifically.

Ogata, *loc. cit.*,[1] states that:

In the Akita area there is a legend that the *Kedani* lives on the bottom of the Omono River and that if a man is bitten by it, he will die. There are old women called *Kedani Baba*, who are said to be able to expel the mites by passing their hands over the body of the person bitten. There are many who go to such a person before visiting a doctor.

The term *Kedani* (hairy-mite or -tick) is the term of preference for the vector in the Akita region, and in about 1909 Hisayoshi Kishida apparently gave the vector mite the scientific name of *Kedania tanakai*; but this name was never published with a description in a manner that would validate it. The earliest valid name is therefore *Trombicula akamushi*, which was given to it by Brumpt in 1910. The genus *Trombicula* is now recognized as comprising diminutive chiggers apparently all parasitizing bats.[1] The vector and its many close relatives are now properly accommodated in the genus *Leptotrombidium*, raised by Nagayo and his colleagues in 1917.

Mites causing scrub-itch seem to have been confused at times with those causing scrub-typhus; also, as is common in the growth of local lore and in the spreading of information from one person to another, genuine observations have doubtless become confused by ignorance or superstition. The earliest written Japanese account of the *tsutsugamushi* itself seems to be that by Seiya Kawakami ('A discussion of the newly-discovered toxic insect') in 1878.[6] His work is discussed below.

[1] See Appendix 1 for a note on this genus. It is a sad fact that had Kishida's new genus *Kedania* been properly published, it would probably have saved some of the later confusion in the classification of the whole family Trombiculidae. As it was, the genus *Trombicula* became for several decades a sort of taxonomic trashcan and in the early stages of research during World War II, the scrub-itch chigger in Australasia, *Eutrombicula wichmanni* was mistakenly identified as *Trombicula minor*, the type species of the genus *Trombicula*, and *E. wichmanni* (under the wrong name of *T. minor*) was mistakenly regarded as the vector of scrub-typhus in New Guinea and Queensland. The disentanglement of this confusion must certainly have slowed down research for a period in that region.

In the beginning of the twentieth century, the infected or potentially infected areas of grassland in the few endemic regions were each tens to hundreds of acres in extent and totalled thousands of acres. The Japanese farmers recognized the considerable localization of the infection by referring to an infected field or typhus-island as *yudokuchi*, which may be translated as 'poisonous place' (*yu*, exist; *doku*, poison; *chi*, land, place). It was well recognized that these *yudokuchi* varied in size and intensity of infection from year to year and that infected areas sometimes moved farther upstream or farther downstream. Endemicity was greatly increased in those years in which heavy winter and spring floods were followed by an unusually hot summer. Sometimes the infection died out and a formerly endemic area became relatively free of infection.

EARLY RECOGNITIONS OF TRANSMISSION OF DISEASES BY ARTHROPODS

Tsutsugamushi disease is not the only example of the naming of a disease by its vector, and of the recognition of arthropod vectors long before this form of transmission was recognized by Western medical scientists. Nor is it the only example where such a local belief met with mistaken 'scientific' criticism, later to be retracted.

In Somaliland, where I first started my tropical medical career, the Somalis have the same word, *kaneo*, for both mosquito and malaria. In the middle of the last century the explorer Richard Burton landed on the coast of British Somaliland and went up into Abyssinia. In more than one of his accounts, he mentions the naïve belief of the savages in those parts that malaria is transmitted by the bites of mosquitoes, whereas it was of course known that it is caused by the bad humours rising from swamps! In other parts of East Africa the local Bantu equivalent of the Kiswahili word *mbu*, mosquito, is similarly used for both mosquito and malaria, which is known as the mosquito-sickness. In all the endemic areas of tick-borne relapsing fever from Somaliland to South Africa, Africans are similarly aware of the connexion between the domiciliated *Ornithodorus* ticks and the infection, even to the extent of knowing that repeated contact with the ticks maintains a relative immunity. The Somalis call both tick and infection by the Arabic word *gurud*. G. A. Walton records that on Ukereri Island in the south of Lake Victoria, the Wakereri refer to malaria as *o'mushyiza gw'emibu* or the mosquito-sickness (*emibu* being akin to *mbu*) and relapsing fever as *o'mushyiza gw'ebibo* or the tick-sickness (*ebibo* being the *Ornithodorus*).[7] My friend Walton tells me of a delightful encounter with the chief of a Wakereri village to whom Walton felt obliged to explain the meaning of a residual insecticide and to tell him that the ticks in the huts of his village could be eradicated for approximately a few cents a head (the East African cent is a hundredth of a shilling). He asked the chief if he wanted to pursue this any further. After a moment's thought, the old chief wisely decided against it. He said that the ticks would surely return again at some later date, at which time it would cost many times more per capita to get rid of them; and at that time the villagers would have lost their immunity and would suffer very much more severely. Also, he went on, his people moved out periodically to neighbouring villages and therefore, again having lost their immunity, and having become infected in these other villages, they would suffer that much more seriously. No epidemiologist or disease control officer could have given a more informed opinion.

It should be a sharp lesson to us that the first result of 'scientific' investigation of tsutsugamushi disease in Japan near the end of the last century led to an attempt to replace the traditional beliefs by a new and fictitious hypothesis. The old belief in transmission by a *mushi* was replaced by a belief in causation by a 'miasma' and the name 'tsutsugamushi disease' was therefore replaced by a new one.

DEVELOPMENT OF THE 'MIASMA HYPOTHESIS'

Scrub-typhus transmitted by *L. akamushi* in northwest Honshu has always been peculiarly distinctive. It was an easily recognized infection with a high mortality. It was readily traced to contact with grassy banks and islets along certain rivers, but only during the summertime. These grassy patches of fertile

Cf, Df, Ef — Coniferous, Mixed Deciduous, Evergreen Forest

FIG. 10. Map of Japan showing geographic restriction of classical scrub-typhus foci (vector *L. akamushi*) to a part of north-west Honshu; an endemic area of 'winter scrub-typhus' (vector *L. scutellaris*) in the Izu Islands; and details of three classical endemic regions and of individual endemic foci or *yudokuchi* ('poisonous places') in Niigata Prefecture. Cf, Df, Ef: Coniferous, Mixed Deciduous, and Evergreen Forest. Adapted from M. Nagayo in A. Byam and R. C. Archibald (1923) *Practice of Medicine in the Tropics*.

silt were attractive to farmers and the disease was obviously occupational.[1] We may assume that some of the more observant farmers saw the mites on the soil and doubtless on their persons

---

[1] Ogata, *loc. cit.*, notes that an infested delta in Niigata was favoured by lovers and distillers of illegal wine. There is a fine spirit of bold enterprise in setting up an illegal still in a *yudokuchi*. As for lovers, love will also always find a place.

while bending down in the fields cutting thatch-grass or clearing and planting. Furthermore, concepts of causation were certainly brought in from the Chinese literature—the fourth and fifth centuries had marked the opening up of Japan, hitherto the secret country, to communication with Korea and China. Also, medicine-men would surely have observed the red mites attached to people, and there must inevitably have come times when a red mite was found attached to a person's skin and was removed, but only to have the characteristic ulcer or eschar develop at the site of attachment, accompanied by the development of fever. (Schüffner describes exactly such an event during an outbreak in Sumatra.) In such cases the medicine-man would have concluded that he had removed the mite too late to prevent the poison from being introduced. The farmers were remarkably precise about their belief, for they recognized that only mites of a particular colour would cause the disease.

Unknown to all these observers, scrub-typhus was in fact present over almost the whole of Japan; but this was completely overlooked until after World War II because it was epidemiologically ill-defined and, in most places, clinically milder. These other forms of scrub-typhus, transmitted by various other species of *Leptotrombidium*, are referred to later. Fig. 10 shows an endemic area of one of these in islands south of Tokyo, and another on Mt. Fuji.

Until the middle of the nineteenth century, Japan had been virtually closed to all western communication. This era of isolationism ended with Commodore Perry's visit in 1853. Following the overthrow of the feudal regime and, later, the advent of the Meiji era in 1869, the gates were opened and there ensued a period of tremendous social and scientific development and the interchange of scholars. Some twenty years after Perry's visit, a German doctor, Erwin Baelz, was given a professorial appointment in Tokyo. This was at a time when the interest of the Japanese civil and military authorities in tsutsugamushi disease was reawakened.

In 1878 and 1879, Tadashi Nagino and the student Seiya Kawakami (writing in Japanese), and Erwin Baelz (writing in German), whilst at first believing that this disease was due to a toxin from the mites, abandoned this farmer's hypothesis in favour of a 'miasma theory'.[1] I have tried to discover how they came to set back the hands of the clock and to produce a nonsensical explanation as the result of the first supposedly scientific investigation of this disease. To produce a mistaken hypothesis is the lot of every scientist from time to time; but to produce it while overthowing the correct hypothesis, and a well-established one to boot, must be the outcome of operations and thought processes that can teach us a lesson.

According to Kawamura,[8] the earliest description of this disease and the 'tsutsuga' was written by Hakuju Hashimoto in 1810. He described conditions in the Niigata Prefecture along the upper tributaries of the Shinano River.

Tadashi Nagino was born in this very Prefecture, and after studying and later teaching Western medicine in Japan, he was appointed in 1871 to Niigata Hospital; and in 1873, after a visit to Tokyo Medical College (which later became part of Tokyo University) in order to study German medicine, he was appointed Chief of the new Nagaoka Hospital in Niigata. In 1877 he was sent temporarily to the fort at Osaka on the outbreak of civil war, and finally retired to private practice in 1889. In 1878[9] he published a brief report on tsutsugamushi disease to Jun Matsumoto, Army Surgeon General, apparently dated December 1877, describing the presence of this disease in a stretch of some 60 kilometres of the Shinano River from Nagaoka to Yokogoshimura in Echigo, Niigata Prefecture. He made recommendations for gathering more information and stated that he had asked the local government to set up a temporary hospital and had invited a foreign doctor (Baelz) to pay a visit and study the infection. He also evidently presented a full account but did not publish this until the next year (25 January 1879) in the *Tokyo Iji Shinshi* (*Tokyo Medical Journal*). [10] This was the first detailed account of tsutsugamushi disease to be published after Hashimoto's report of 1810, a copy of which I have been unable to obtain.[2]

[1] At that time, malaria was ascribed to a miasma, but nobody could have been scientifically confident that he knew exactly what this meant.

[2] See Appendix 7 for translated excerpts from early publications of Nagino and Kawakami (which Ogata, *loc. cit.*,[1] quotes verbatim) as well as some excerpts from Ogata's historical review.

Erwin Baelz was born in Bietigheim in southern Germany and graduated in medicine from Tübingen University. In 1876 at the age of only 26 he was invited to be a Professor of Internal Medicine at the University of Tokyo, where he remained until 1901. Baelz married in Japan and lived there until he was 52. His son, Toku (Tokunosuke) was a student of Japanese drama and attempted to introduce Japanese *Kabuki* to Germany. Toku edited his father's diary.[11] Baelz travelled widely during his first year in Japan and at that time was introduced to tsutsugamushi disease in Niigata. In 1878, apparently at the invitation of Nagino, he went back to Niigata for a brief investigation, accompanied by a medical student Seiya Kawakami.

At this time, a medical missionary in Niigata, Theodore A. Palm, wrote a letter to a Rev John Lowe, which was published in the *Edinburgh Medical Journal*, 1878, as 'Some Account of a Disease called "Shima-Mushi" or "Island-Insect Disease", by the Natives of Japan; peculiar, it is believed, to that country, and hitherto not described.'[12] He was taken, on 1 August 1878 (?), to a Buddhist temple about two miles from Nagaoka, converted into a temporary hospital for tsutsugamushi disease patients. He mentions finding 'that an intelligent young man, a student from the medical school in Tokio, conducted by German professors, had been appointed as house-surgeon'. This must have been Seiya Kawakami. Palm stated that

> Later on in the month of August, the house-surgeon, above-mentioned, called on me and stated that he had discovered the insect, that while picking with a needle the black spot in the centre of the swelling of a patient recently affected, he noticed moving up his finger a very minute insect, which, upon microscopic examination, resembled a spider. To preserve his specimen he had stuck it with gum between two glass slides, and offered it to me for examination. His method of preserving it, however, had rendered it impossible to do more than recognize that it was an insect of some kind, but it may have been an *Acarus scabiei*. Under the circumstances, I do not consider it yet demonstrated that it is caused by an insect.

Palm was correct in doubting that this creature could have been a *tsutsugamushi* abandoning the eschar. I quote this passage as an example of the danger of drawing hasty conclusions under the intoxication of such enthusiasm as Kawakami's.

Seiya Kawakami was a student at the Tokyo Medical College and spent his summer vacation in 1877 inspecting *yudokuchi* near his home. According to Baelz, it was Kawakami's report of this that so interested Baelz that he decided to visit the infected places with Kawakami. This he did in August 1878, but Kawakami may have stayed there for a longer period. Palm mentions a person who must be Kawakami but makes no mention of Baelz. As a result of this work, Kawakami published several reports in Japanese, two of these being important contributions which Ogata, *loc. cit.*, has quoted directly; and Baelz published a long account in German in Virchow's Archives, citing Kawakami as a collaborator.[13]

Kawakami's first contribution was published in 1878.[6] Having quoted the local belief, he shows signs of doubt. The question at issue for the historian is who first sowed the seeds of doubt, or at least encouraged disbelief most actively, among the trio, Nagino, Baelz, and Kawakami. A translation of part of Kawakami's text reproduced by Ogata reads as follows:

According to the inhabitants of the area, the disease is caused by the bite of so-called *tsutsuga no mushi* or *shima mushi*. The environments in which people are bitten and infected are usually found on islands and along the banks of rivers. . . . When we think of it, it occurs to us that we have never seen any *tsutsuga no mushi*, nor have others seen them. The villagers and local medicine-men say that the *tsutsuga no mushi* is about three centimetres long and looks like a thin hair.[1] They are sometimes white and sometimes brown in colour. When they are burned, they give off an odour. Some villagers say that there are two kinds of these insects, one white or brown, the other red. The former does more damage than the latter and causes systemic symptoms to occur when a person is bitten. . . . I observed the insect under the microscope and saw baby insects come out of the large one and move around. [Ogata: This is doubtful.] In

[1] These remarks about the resemblance to a hair, and the odour when burned, are also repeated in the second paper, Ogata adding in parenthesis that in the previous year, Seiya placed a specimen under the microscope and found that it really was a hair! No wonder it had been reported that 'they' give off an odour when burned —it would have been an odour of burned hair! This test is surely the most astonishing ever to have been applied to an arthropod—could it have arisen through plucking out hairs to destroy crab-lice or eggs (nits)? It is known that infected fields were burnt as an effort at control, but burning individual mites is not mentioned.

three days the baby insects changed into small insects with four legs. [This is too short a time for metamorphosis of chiggers.] Subsequently, they were lost to observation through the playing of the children. [This remark conjures up a vivid picture and the sympathy of all research workers.]

Kawakami then goes on to note that the person does not feel any sensation at the time of the bite (except that touching an attached mite will give a pricking sensation), the lymph glands subsequently swell and fever develops. His clinical descriptions are very good. His statement about the white insects causing systemic symptoms is contradicted in his second paper, where the red ones are dangerous: this must have been a slip of the pen.[1] His observations on the baby mites are wholly inapplicable to chiggers. His drawing of the six-legged larva could be one of many kinds of trombidiform mite, but his drawing of the eight-legged form, either nymph or adult, is utterly unlike a chigger although it does resemble other forms of red trombidiform mites which are very common in soil. Kawakami was evidently looking for mites from the ground and did not inspect any of the field-mice or voles. I cannot help wondering if he did not adopt the role of the laboratory scientists, having material brought to him from the field, under the influence of a very reasonable fear of intensively studying the dangerous places at first hand.

[1] Kawakami also suggests that the *akamushi* (red-insect) is so named because 'when the insect makes a hole in the skin and sucks blood, it becomes red and visible'. This observation has been independently made by others and quoted in the literature. It is an example of hasty observation without thought. The unfed larva is already bright red or orange before it feeds, and indeed it becomes distinctly paler with engorgement, Moreover, this red colour is common amongst the post-larval stages which do not feed on animals at all. It is also the characteristic colour of trombidiform mites generally, the pigment apparently being related to that of carrots. Larval chiggers feed on tissue-juices imbibed through their drinking-straws or stylostomes. After this became known, it became fashionable amongst entomologists to be derisive when anyone suggested that chiggers fed on blood. Nevertheless, the guts of engorged chiggers do often contain blood-pigment and Kepka[14] has photographed ingested blood-cells. Vercammen-Grandjean, however, noted that when larvae of species of *Schoutedenichia* became faintly pink through imbibing blood, they failed to develop into nymphs.[15] A contrary situation has obtained with ticks: workers obstinately supposed them to imbibe only blood until it was found that the larvae of at least one species feed only on tissue-juices in much the same manner as trombiculids but without the drinking-straws.

Medical entomology was unheard of in those days and Kawakami and the others after him had nothing to guide them in their studies of these tiny mites which demand the highest powers of the microscope. Indeed, Kawakami seems to have overlooked the need for an animal host to maintain the mites, just as Gilbert White seems to have overlooked the significance of the rabbit.

In his second more thorough paper published on 15 March 1879,[16] Kawakami has abandoned the idea of transmission by chiggers and he entitles the paper *Suison-netsu* (flood-fever), referring in his first paragraph to the confusion hitherto prevailing in the name of the disease, as caused by the bite of a toxic insect. 'I have called it *kozui-netsu* [also meaning flood-fever] or "Überschwemmungsfieber" . . .'—but so had Nagino. [10] He then marshals his negative evidence: (a) 'We do not know the species to which the *tsutsuga no mushi* belongs, nor have we seen it.' (He then repeats the story about its looking like a piece of hair and giving off an odour, but he also says that the 'so-called *akamushi* is very tiny, barely visible to the unaided eye'.) (b) 'Last year, I found some insects that I thought were toxic, but these were just plain ticks that I believe to be harmless to humans. If the disease is caused by these insects, those of us in the hospital should have been infected but we were not.' Is this evidence that he examined only specimens that had been brought to him in the hospital? (c) The evidence of both patients and doctors seemed wholly unreliable and not based on facts. (d) 'The villagers do not believe that the bite of the *akamushi* is the cause of the disease, because they have been bitten by as many as six or eight of the red insects and have not experienced any infection. . . . A patient was sent to us this year from Nagaoka hospital. He had been bitten by akamushi. We removed the insect. The patient did not become infected.'

He concludes that the disease is caused by some sort of chemical toxin and not by the bite of the *tsutsuga no mushi* or *akamushi*. 'Such a chemical may be something produced in the ground from the accumulation of organic matter at the time of the spring floods or from spoilage due to heat in the summer.'

Baelz and Kawakami published an account of 'Fluss- oder

Überschwemmungsfieber' in Virchow's Archives in September 1879.[1] Later in the same volume is a note by Baelz identifying Palm's island-mite fever with Baelz's Japanese river-fever.[17] Although these two papers did much more than Palm's publication in a 'local' journal to draw the attention of Western medicine to this interesting disease, they contain hardly more than can be found in the earlier papers by Nagino and by Kawakami. Indeed, Ogata maintains that the paper by Baelz and Kawakami is merely a German translation of the Japanese. This is not wholly just: there are differences and the data are better marshalled.

What happened to Kawakami after 1879 is a mystery. He published no more[2] nor was he later listed among the graduates of Tokyo University.[1] Did he succumb to the infection himself as several other investigators have done in the past? This would however have put him on a kind of roll of honour and would probably have been recorded. Did he sink into oblivion after rendering himself unpopular, a mere student[3] in the company of the two professors, by being carried away by too much enthusiasm after being given what was probably the main responsibility for the entomological, and therefore epidemiological, investigations? As the expression goes, he was the lowest man on the totem pole; and as such should have been seen but not heard.

It is very difficult at this late date to get a clear picture of the interactions between these personalities and to sort out the

[1] Baelz's diary entry for 21 September 1879: 'At long last I have finished my papers on Japanese river fever, based upon my study of the disease last year in Echigo. I posted it this evening to Virchow in Berlin for publication in his "Archives".'[11] It is a curious fact that either the diary never mentioned events in Echigo or these have been edited out. A mistake had been made, and the tsutsugamushi was soon accepted again as the vector and the name tsutsugamushi-disease restored. By his stern Germanic upbringing, Baelz would have had great difficulty in admitting that he had erred. We may safely assume that he never wanted this subject brought up again.

[2] Ogata, *loc. cit.*, notes two reports by Seiya Kawakami and Masahashi Kawakami, 'On a kind of fever in the Nagaoka area' in *Hokuetsu Gakkai Kaiho* (*Rep. Hokuetsu med. Soc.*) No. 72; and *Iji Hoshu* (*Outstanding Reports in Medicine*), Nos. 26, 29. I have not been able to secure these.

[3] Ogata, *loc. cit.*, states that although Baelz headed his report as being in conjunction with the medical student, Kawakami, the latter 'should not be regarded as co-author'. Did Baelz include him because he had drawn freely from Kawakami's publications?

contributions which each made. One may take account only of the ages and relative experience of Nagino, Baelz, and Kawakami, of national characters, and of the dates of publications and material contained therein. Nagino was the senior man who would have been familiar with the lore of tsutsugamushi in his home county or prefecture, even from the time before he was a medical student; and he would certainly have been seeing cases in one or another hospital from 1871 onwards. According to Ogata, a preliminary report by Nagino on the disease appears in the very first issue of the *Tokyo Iji Shinshi* published in February 1877.

Baelz was a young and undoubtedly eager physician with a new appointment in Japan, without doubt stimulated by this very charming country and people and fascinated by the new medical problems confronting him. After his first visit to Niigata in 1876, it is clear that Nagino invited him to come back to investigate tsutsugamushi disease in collaboration. It might be significant that Baelz does not mention Nagino. There could readily have been a clash of individualistic personalities. Kawakami, the youngest, and still a medical student, had one advantage over Nagino in that he had performed an autopsy on a tsutsugamushi patient (whereas Nagino could get no permission for autopsy on patients whom he had seen) and had several advantages over Baelz, namely this experience and also his familiarity with the language and the customs. Since many educated Japanese spoke German, Baelz could get along in that language and thus may have been slowed up in mastery of the difficult Japanese language; in any case it would probably have taken him a very long time to be able to converse freely with the farmers. Nagino would have been the chief of a very busy hospital, definitely a local medical authority on tsutsugamushi disease, and, on the whole, would have tended to be steeped in local tradition. By comparison, I imagine Baelz was probably more imbued with the somewhat impatient, didactic scientific confidence which persists even to this day, a confidence that things are black or white but never vaguely grey. He would also have been rather less impressed by local traditions and the beliefs of non-scientific farmers. He might indeed, like others before and after him, have considered it preposterous that an arthropod could transmit an infectious disease, for we must remember this was 21 years before Ronald Ross, followed by Grassi and Bignami, demonstrated that mosquitoes transmitted malaria. Incidentally, these early writings on tsutsugamushi do not mention the existence of temples or shrines dedicated to placating the vector: these delightful facts seem to have been discovered decades later by workers on this disease.

Reading the reports of Nagino, Baelz, and Kawakami, I cannot help visualizing Kawakami as the eager young man doing most of the field-work and the contacting of people with local lore and experience, learning much but at the same time becoming confused (as even an experienced entomologist might have done in the circumstances) by the presence of mites of several different species, each species existing in two quite distinct forms (larva and adult), and by the presence of a proportion of individuals (perhaps 90 per cent or more) of the true vector being uninfected. In addition he must have been bewildered by the statements of the local doctor and some residents that the 'insect' looked like a piece of hair. Kawakami states quite clearly in one place[16] that in spite of the long-standing belief that the bite of a tsutsugamushi caused the disease, the villagers he was dealing with did not believe in this because they had been bitten by as many as eight of the red mites and had not suffered any infection. It is however uncertain how much Kawakami may have influenced the answers by leading questions.

The villagers themselves were undoubtedly resentful of these bustling interlopers and specimens may have been brought in either mischievously or by simpletons. Baelz[13] says:

The behaviour of the people makes exact studies very difficult ... we found resistance, almost conspiracy ... Though public officials supported us magnificently, and though we had a big airy room for free-of-charge treatment of patients for two weeks not a single patient came to Mr. Kawakami for admission ... Patients visited in their filthy homes [Baelz reveals much of himself in his writing] could not be moved either to seek treatment or to accept any medicine of European origin ... for fear of autopsy ... We had no choice but to ask for help from the police ... succeeded in transporting patients to hospital ...

We find this trio of early workers in these circumstances putting aside the traditional belief in the transmission of tsutsugamushi disease and maintaining that it was caused by a miasma. Nagino shows some conversion to this idea in his 1879 paper, whilst Kawakami shows considerable vacillation, all the time investigating the local chiggers more and more. Local field investigations at that time showed that the situation was not nearly so simple as it may have seemed. The villagers in that particular locality may have become doubtful of tradition. I suspect that Kawakami, trained to think in terms of sick people, utterly inexperienced in epidemiological field-work, and presented with an exceptionally difficult problem in medical entomology, succeeded only in confusing matters; and that Baelz in particular, as the immigrant scientific sceptic, crystallized the growing doubts in favour of what later workers have called 'Baelz's miasma theory'.

SUBSEQUENT SCIENTIFIC INVESTIGATIONS IN JAPAN

Subsequent investigations in Japan have been competently reviewed by a number of authors[1] and I shall mention only the highlights. For some decades the major centre for the study of tsutsugamushi disease was the Institute for Infectious Disease founded in Tokyo in 1893 by Kitasato who had returned from Germany after studying bacteriology under Robert Koch. This Institute later published the *Kitasato Archives of Experimental Medicine* with papers in English and German. The first issue in April 1917 contained a paper on the life-cycle of the *akamushi* by Miyajima and Okumura.

Keisuke Tanaka in 1899 reported his studies in Akita Prefecture and clearly described the tsutsugamushi as a bright red larva.[18] This important paper reinstated tsutsugamushi disease or kedani-fever as an arthropod-borne disease not caused by a miasma. Later, Nagayo and his colleagues in Yamagata Prefecture made the greatest single advance in the taxonomy of the trombiculids when in 1921 they described five species and proposed the genus *Leptotrombidium*.[19]

Claims to the recovery and identification of the causative

---

[1] See Appendix 5 for details.

AKAMUSHI: THE RED MITES OF JAPAN

agent of tsutsugamushi disease were meanwhile being made, five names having been proposed by 1931, with the specific names *tsutsugamushi* Hayashi, *nipponica* Sellards, *orientalis* Nagayo et al., *tsutsugamushi* Ogata, and *akamushi* Kawamura and Imagawa.[20–6] There is at present some difference of opinion on the correct name, either *tsutsugamushi* (Hayashi) Ogata (or *tsutsugamushi* Ogata as Ogata would have it)—on the grounds that it cannot be proved that Hayashi never did have the true causative agent, even though he may have had more than one species—or *R. orientalis* Nagayo et al., if Hayashi's claim is invalidated. Most workers in Japan seem to accept the latter name, but Bergey's Manual accepts *tsutsugamushi*. By far the most heated controversy among investigators has concerned being the first to recover the causal agent. As Ogata says, this is like a gamble: one investigator wins, and all the others lose in spite of their competence, eminence, and many years of effort. It is true that there is often a grossly unfair element in the apportionment of kudos in such matters. Some would regard the Nobel Prize in this light. Not only is one's self-esteem involved, but academic advancement and the opportunity to do more research. Let us be both generous and just in such a circumstance and at least share in our minds the credit for achievement among these devoted investigators.

Formosa was investigated successively by Hatori, Kawamura and Yamaguchi, and Morishita.[2, 27, 28] The vector, *L. akamushi*, in that island was widespread, occurring in places away from rivers and on a great variety of hosts. The infection was milder with a lower mortality rate than in Japan. The Pescadores (Hoko) Islands, between Formosa and the mainland were found by Kawamura and Yamamiya[29] to harbour domestic infection in people's gardens. The incidence was highest among children and young people and the mortality rate was very low. These islands have recently been investigated again thoroughly.[30]

It is curious that in all this time practically nothing was reported of this disease in China, whence the earliest reports of serious illness brought on by the 'sand-mite' had emanated centuries before.

During World War II the Allied forces recorded well over

16 000 cases of scrub-typhus. There was a commensurate number in the Japanese forces. The momentum of wartime research on scrub-typhus actually increased after the war especially in Japan, in Malaya, and in some laboratories in the USA. Some of this is briefly noted in Chapter 2 (see also Appendix 5). In his review of wartime studies of scrub-typhus by members of the United States Army, Philip has given information on investigations in Japan after the war.[31]

RIVALRY, REASON, AND RESEARCH

The creative activity called research demands much energy, devotion, and time for its pursuit. The true devotee always seems to find the time; nothing can hold him back. Among the energy-giving processes that contribute to the drive of the investigator are those that arise from within—the burning fever to work out one's thoughts and the desire to achieve prestige or simply recognition—and those that are imposed from without, such as the encouragement or rivalry of others and the need to achieve academic advancement. I have in these lectures tried to bring out the obstacles to the free thinking necessary in research, such as the effects of tradition and, sometimes, the 'first prejudice' formed at the time of one's first incomplete glimpse of a situation. It is therefore pleasant to consider some of the helpful stimuli from without. The subject of creativity from within has been dealt with by different people in many different ways.[32, 33]

Ogata in his musings (*loc. cit.*) has the following to say on rivalry as a stimulus in research:

Tsutsugamushi disease has been the topic of research in Japanese medicine since 1877. There has probably been nothing else that kept the scientific meetings and journals bustling so much for over half a century.... Looking back over the entire history of this research, we find a competitive or even hostile spirit dominant among the researchers of various academic cliques. Whenever tsutsugamushi was the topic at meetings, discussions were numerous and vigorous. At times there were riotous scenes of personal attacks and mudslinging [*doroshiai*, literally mud-contests] which attracted the attention of the public.

Research is centred around personalities, not around things. When people are engaged in the study of the same subject, competition arises. On all sides there are efforts to do the impossible or the rash in research. The spurring of efforts results in progress. Progress and development are slower when there is no academic opponent.

In the study of the pathogenic agent of tsutsugamushi disease, Ogata, Nagayo, Kawamura, and Hayashi pushed in the direction of a rickettsial cause and this was shortly verified. Decisive results are more quickly obtained when there are contending investigators.

One cannot play *go* or *shogi* [Japanese chess] without an opponent. While playing, the opponents may get angry at each other and may feel the urge to knock each other down, but they cannot enjoy the game without each other. If they are separated, they will begin to miss each other. The same thing applies to competition and rivalry in research. One feels lonely without opponents.

Disagreement, competition, and jealousy are among the components of rivalry. Disagreement is expressed by criticism, competition by secrecy and various forms of opposition, jealousy by obstruction and vilification.

Criticism, even though unfriendly, is a splendid corrective for the investigator. It forces him to sharpen his arguments, think through his problems and plans, straighten out his perspective. It is surprising how vague some of a person's thoughts may be until he is taxed by a critic. But it takes time and self-control to withstand the insult of criticism while one digests and acts on it. The more one builds up a system of thought, the child of one's mind, the more one cherishes it. The first reaction to disagreement is some degree of resentment and sometimes this is violent, directed against the critic rather than the circumstance. It may lead to unexpectedly displaced responses. Several years ago while lecturing to a gathering of Russian parasitologists in the USSR, I outlined G. A. Walton's work and his hypothesis that, although louse-borne relapsing fever has evidently evolved from tick-borne forms of spirochaetes, evidence in East Africa could well be interpreted to imply that the human louse-borne strains had been conveyed by man to the domiciliated soft-tick and been duly modified in that new host. (This was supported by the apparent absence of a wild reservoir of the tick-borne infection which seemed to be strictly human and domestic.) Two elderly parasitologists in the audience who had specialized on spirochaetes were so offended by this heresy that they stated that such a ridiculous adaptation would be impossible and both accused me of 'denying the fact of evolution', although I was, in fact, suggesting that evolution did not stop with Darwin's death. Several Russian scientists afterwards explained to me apologetically that people sometimes get fixed ideas and resent new ones. The wise scientist realizes that the more he is obliged to change his ideas the closer he is approaching the truth. Each hypothesis is built on the residues and wrecks of others. But the scientist who is critical of another should be considerate of his victim's brain-child. Even the harshest criticism can be kindly, even concealing such unworthy emotions as jealousy if they are indeed present.

Scientific jealousy has been with us for as long as scientific thought. Nowadays it is greatly encouraged by the spirit of competition rife in both the academic and industrial worlds.[1] 'Publish or perish' is the saying in academic circles. Scientists may find themselves involved willy-nilly in rat-races for priorities. Unfortunately, this is often most intense in the earlier years when the scientist is establishing not only his reputation but his habits of thought.

It is a curious fact that an outstanding achievement in literature or art is regarded as an unique product of the artist, but many highly creative achievements in science either are, or can more readily be, regarded as fruit plucked because that particular scientist got there first—possibly more through chance or availability of funds than because he towered over his colleagues in creativity. The chief reason for this is that new vistas in science are seen by standing on the shoulders of those who have gone before; and there repeatedly comes a time when scientific advances create a platform from which scientists can see new vistas (through as many different eyes) and prepare

---

[1] Baelz's diary[11] has this sole entry for 31 January 1879: 'Today I am thirty years old. Thirty! Yet so far I have made no contribution to science. This alarms me; and it is all the worse seeing that I have stored up abundant materials which urgently demand publication.' Note the nature of his concern: unpublished research is no contribution to science—it remains as something for one's private edification; but it becomes irksome because it is unrecognized by others.

their expeditions to explore them. These expeditions may then comprise lone explorers or large teams, scientists with limited resources or others with ample funds and all modern facilities. Thus it may seem that final achievement is more related to who arrived first than to outstanding creative ability. This attitude is of course unfair. The achievement may give credit to several people for superb creativity, but who reaches a particular goal first is often simply a gamble among the elite. Incidentally, the scientific achievements that ensue do not usually meet with wide and immediate public appreciation because they are hard to understand. With regard to literature, perhaps outstanding achievements do in fact resemble scientific achievements more often than one would think: the stage is set, by current fashions in thought and manners, for further light to be thrown on man's past or present work. A sufficiently attractive exposé meets with wide and immediate public recognition. To take an example of a literary effort in a scientific field, Rachel Carson[34] very successfully scooped Robert Rudd,[35] and presumably several others, by a combination of getting there first and succeeding in giving *Silent Spring* much more popular appeal than Rudd's more scientific *Pesticides and the Living Landscape*; yet the latter was probably almost ready for press before the former.

Rival thinkers gather rival schools of thought. What we have failed to learn from history is that in such circumstances each contender often has an essential part of the whole truth. The danger of such forms of rivalry is that the emotions aroused tend to close one's eyes and ears to the other person's postulates. Emotions are like electrostatic charges, the separate proponents build up equal charges and are driven apart, while the truth may lie in common ground between them. A classic example of this was the rivalry over the undulatory versus the corpuscular theory of light, resolved when it was realized that light consisted of packets of waves and could behave in one way or another according to the conditions of observation. In modern science, many opposing schools have developed, for example, in explaining the factors controlling animal populations—one's 'first prejudice' in this field tends to be decided by the species of animal on which one happens to start work.

There is similar rivalry in theorizing about biological clocks. Such rivalry is stimulating and salutary; but when criticism becomes captious, when issues become more personal than scientific, and when a necessary meeting of minds seems unattainable because of these, then it is time for a referee to bring about reconciliation.

Something different is involved in what we have referred to as a sort of gamble, such as Ogata commented on. There arise many occasions when simply time, opportunity, and luck decide which of a set of competent investigators or teams produces the right answer first. The blow to the losers may be disproportionately severe if we take account of the competence, energy, and dedication of all the contenders. It may be the discovery and naming of the causal agent of scrub-typhus, or the obtaining of an acceptable poliomyelitis vaccine. There is little doubt that the emergence of an effective killed-virus polio vaccine held back the development of what was recognized as the ideal goal, the live-virus vaccine. Finally, at least two eminent workers and their colleagues emerged as losers after tremendous investments.

Ogata seems to have welcomed even hostile rivalry, almost with a glint of battle in his eye. A Western scientist might readily misinterpret the translated statements of a prominent Japanese scientist who graduated in medicine from Tokyo University in 1916, and who, as we have seen, belongs to a family devoted to tsutsugamushi research. Even those of us foreigners who have had the pleasure of seeing something of the Japanese in their own country may have great difficulty in gaining genuine insight. For example, one aspect of personality that would concern us now was summarized thus by the psychiatrist Takeo Doi[36]:

... He [Hajime Nakamura[37]] says that the Japanese way of thinking is greatly influenced by an emphasis on immediate personal relations and also that the Japanese have always been eager to adopt foreign cultural influences, but always within the framework of this emphasis on personal relations. To state this in psychoanalytic terms: the Japanese are always prepared to identify themselves with, or introject, an outside force, to the exclusion of other ways of coping with it. This character trait of the Japanese was touched upon by Benedict,[38] too, when she said that 'the Japanese have an ethic of alternatives' and 'Japan's motivations are situational' . . .

It may be that the Japanese scientist makes demands and has needs for the stimulation of both his creativity and his industry, and that these are not the same for most Western scientists—except perhaps that a German might come closest to understanding. Doi also stresses the importance to the Japanese of the concept of *amaeru* (to depend and presume upon another's benevolence). Without being a psychiatrist I cannot help feeling that the urge to *amaeru* is associated with the exceptionally close communion the Japanese have with nature, feeling much more an integral part of it than can readily be appreciated by a Westerner. I also suspect that a deep urge to *amaeru*, but in a more childlike way without the roughness of rivalry, served to stimulate Gilbert White but at the same time to stop him from going further. I feel sure that there would never have been a *Natural History of Selborne* had it not been for constant stimuli from White's mentors, Pennant, Hales, Barrington and the others. In 1767 he wrote the following which should be compared with Ogata's sentiments:

It has been my misfortune never to have had any neighbours whose studies have led them towards the pursuit of natural knowledge; so that, for want of a companion to quicken my industry and sharpen my attention, I have made but slender progress in a kind of information to which I have been attracted from my childhood.[39]

Rivalry, whether healthy or unhealthy, always acts as a splendid spur. Yet I suggest that research or let us say creativity, progresses more rapidly through friendly criticism and endless discussion among colleagues who are not establishing rival camps. Animosity reflects on the maturity of a scientist, and I am sure that the best interests of careful scientific research are seldom served by an atmosphere of hostility. The development of the urge from within and education by constant demands to clarify and justify are better than the spur of hostile rivalry from without. Scientists come in all shapes, however, and there are those who would thrive on fighting. I would suspect, nevertheless, that he who requires hostile rivalry is himself a hostile rival waiting for an opponent. As a spur from without, rivalry is benign compared with demands to produce and the ensuing spirit of commercial competition. Those who provide large-scale support for research have these and many other difficult problems to think about. Theirs is an appalling responsibility.

Since these lectures went to press, I have found an account by the sociologist Robert K. Merton on scientific priorities and the various factors that may lead scientists into various forms of deviant behaviour: Priorities in scientific discovery: a chapter in the sociology of science. *Amer. Sociol. Rev.* **22**, 635–59, 1957.

REFERENCES

1 OGATA, N. (1953). [Seventy-seven year history of research on tsutsugamushi disease in Japan] (in Japanese). *Tokyo Iji Shinshi* [*Tokyo med. J.*] **70**, 51–73.
2 HATORI, J. (1919). On the endemic tsutsugamushi disease of Formosa. *Ann. trop. Med. Parasit.* **13**, 233–58.
3 TAMIYA, T., Editor (1962). *Recent Advances in Studies of Tsutsugamushi Disease in Japan.* Tokyo, Medical Culture Inc., 308 pp.
4 KAKURAI, T. (July 1915). [Tsutsugamushi disease in Yamagata Prefecture] (in Japanese). Tokyo Printing Co., for Police Department of Yamagata Prefecture. ('Not for sale'). 58 pages. (See Appendix 7.)
5 SAMBON, L. (1928). The parasitic acarians of animals and the part they play in the causation of the eruptive fevers and other disease of man. Preliminary considerations based upon an ecological study of typhus fevers. *Ann. trop. Med. Parasit.* **22**, 67–132.
6 KAWAKAMI, S. (1878). [A discussion of a newly discovered toxic insect] (in Japanese). *Tokyo Iji Shinshi*, Nos. 20, 21. [Quoted from OGATA, [1]. Note: Seiya, not Seijo nor Motosaburo Kawakami.]
7 WALTON, G. A. (1962). Notification of the replacing of the single species concept of '*Ornithodorus moubata*' by a group of new species and the creation of a neotype for the nomen nudum *moubata* Murray, 1877. (Correspondence.) *Trans. roy. Soc. trop. Med. Hyg.* **56**, 91–2.
8 KAWAMURA, R. (1926). Studies on tsutsugamushi disease (Japanese flood fever). *Med. Bull., Coll. Med., Univ. Cincinnati,* **4** (Special Nos. 1, 2), 1–229.
9 NAGINO, T. (1878). [Record of the Army General Hospital. A Memorial] (in Japanese). *Tokyo Iji Shinshi* [*Tokyo med. J.*] **20** [Quoted from OGATA[1]]
10 NAGINO, T. (1879). [Summary of hospital records on tsutsugamushi disease] (in Japanese). *Ibid.* **40**(4) [Quoted from OGATA[1]]

11 BAELZ, T., Editor. Transl. by E. and C. PAUL (1932). *Awakening Japan: the Diary of a German Doctor: Erwin Baelz.* New York, Viking Press.
12 PALM, T. (1878). Some account of a disease called 'shima-mushi' or 'island-insect disease' by the natives of Japan; peculiar, it is believed, to that country, and hitherto not described. *Edinb. med. J.* **24**(pt. 1): 128–9.
13 BAELZ, E. and KAWAKAMI, S. (1879). Das japanische Fluss- oder Überschwemmungsfieber, eine acute Infectionskrankheit. *Virchow's Arch.* **78**, 373–420.
14 KEPKA, O. (1964). Die Trombiculinae (Acarina, Trombiculidae) in Österreich. *Z. Parasitenk.* **23**, 548–642.
15 VERCAMMEN-GRANDJEAN, P. H. (1958). Revision du genre Schoutedenichia Jad. et Verc. *Ann. Mus. roy. Congo belge, Tervuren* (Sci. Zool.) **65**, 7–101, p. 13.
16 KAWAKAMI, S. (1879). [A second report on the toxic insect: flood fever] (in Japanese). *Tokyo Iji Shinshi* **51** [Quoted from OGATA[1]]
17 BAELZ, E. (1879). Nachtrag zu dem Aufsatz über Flussfieber. *Virchow's Archiv.* **78**, 528–9.
18 TANAKA, K. (1899). Über Aetiologie und Pathogenese der Kedani-Krankheit. *Zbl. Bakt., I. Abt. Orig.* **26**, 432–9.
19 NAGAYO, M., MIYAGAWA, Y., MITAMURA, T., TAMIYA, T. and TENJIN, S. (1929). Five species of tsutsugamushi (the carrier of Japanese river fever) and their relation to the tsutsugamushi disease. *Amer. J. Hyg.* **1**, 569–91.
20 HAYASHI, N. (1920). Etiology of tsutsugamushi disease. *J. Parasit.* **7**, 53–68.
21 SELLARDS, A. W. (1923). The cultivation of a rickettsia-like microorganism from tsutsugamushi disease. *Amer. J. trop. Med.* **3**, 529–46.
22 NAGAYO, M., TAMIYA, T., IMAMURA, A., SATO, K., MIYAGAWA, Y. and MITAMURA, T. (1924). Demonstration of the virus of tsutsugamushi disease. *Trans. Jap. path. Soc.* **14**, 193.
23 NAGAYO, M., TAMIYA, T., MITAMURA, T. and SATO, K. (1930). Sur le virus de la maladie de tsutsugamushi. *C. R. Soc. Biol.* **104**, 637–41.
24 NAGAYO, M., TAMIYA, T., MITAMURA, T. and SATO, K. (1930). On the virus of tsutsugamushi disease and its demonstration by a new method. *Jap. J. exp. Med.* **8**, 309–18.
25 OGATA, N. (1931). Aetiologie der Tsutsugamushikrankheit: Rickettsia tsutsugamushi. *Zbl. Bakt., I Abt. Orig.* **122**, 249–53. (See also *Trans. 8th Congr. Far Eastern Association trop. Med.*, Siam, December *1930* **2**, 176, published June 1932.)
26 KAWAMURA, R. and IMAGAWA, Y. (1931). Die Festellung des Erreger, bei der Tsutsugamushi-Krankheit. *Zbl. Bakt., I Abt. Orig.* **122**, 253–61.
27 KAWAMURA, R. and YAMAGUCHI, M. (1921). Über die Tsutsugamushi-Krankheit in Formosa, zugleich eine vergleichende Studie derselben mit der in Nordjapan. *Kitasato Arch. exp. Med.* **4**, 169–206.
28 MORISHITA, K. (1942). Tsutsugamushi disease: its epidemiology in

Formosa. *Proc. 6th Pacific Sci. Congress, California, 1939* **5**, 639–47. Berkeley, University of California Press.
29 KAWAMURA, R. and YAMAMIYA, C. (1939). On the tsutsugamushi disease in the Pescadores. *Kitasato Arch.* **14**, 79–91.
30 COOPER, W. C., LIEN, J. C., HSU, S. H. and CHEN, W. F. (1964). Scrub typhus in the Pescadores islands: an epidemiologic and clinical study. *Amer. J. trop. Med. Hyg.* **13**, 833–8.
31 PHILIP, C. B. (1964). Scrub typhus and scrub itch. Chapter 11, pp. 275–347 in *Preventive Medicine in World War II*, Vol. VII, *Communicable Diseases. Arthropodborne Diseases Other Than Malaria.* Washington, D.C., Office of Surgeon General, Department of Army.
32 TAYLOR, C. W. and BARRON, F., Editors (1963). *Scientific Creativity: Its Recognition and Development.* Utah Creativity Research Conference. New York and London, John Wiley, 419 pp.
33 GHISELIN, B., Editor (1952). *The Creative Process. A Symposium.* Berkeley and Los Angeles, University of California Press. [Actually an anthology.]
34 CARSON, R. (1962). *Silent Spring.* Boston, Houghton Mifflin Co. [Also in paperbacks.]
35 RUDD, R. L. (1964). *Pesticides and the Living Landscape.* Madison, University of Wisconsin Press.
36 DOI, L. T. (1962). *Amae:* A key concept for understanding Japanese personality structure, pp. 132–9 *in* R. J. SMITH and R. K. BEARDSLEY, Editors, *Japanese Culture: Its Development and Characteristics.* Chicago, Aldine Publishing Co., 193 pp.
37 NAKAMURA, H. (1960). *Ways of Thinking of Eastern Peoples.* Tokyo, Japanese National Commission for UNESCO (Comp.), Japanese Govt. Printing Bureau.
38 BENEDICT, R. (1961). *The Chrysanthemum and the Sword.* Boston, Houghton Mifflin Co.
39 WHITE, G. (1947). *The Natural History of Selborne.* Letter X to Pennant, 4 May 1767. London, Cresset Press.

3

EMERGENCE OF THE TYPHUS GROUP OF FEVERS

THE name *typhus* comes from a Greek word meaning 'stupor'. This is a doubly appropriate name for not only is stupor a striking and characteristic feature of the toxaemia of typhus and other fevers which have been confused with it, but the Greek word *typhos* also means smoke or haze and, until fairly recent times, the complex of diseases related to typhus was but darkly seen through the smoky clouds of our ignorance.

In the beginning of the eighteenth century, three totally different fevers, each characterized by a toxaemic stupor, were being confused under the same name, typhus, which was given to them by Sauvages in 1760.

Willis and Sydenham were the first to distinguish these fevers but it was not until the mid-nineteenth century that the three were quite clearly differentiated into *typhus abdominalis* (typhoid or enteric fever), *typhus recurrens* (relapsing fever), and *typhus exanthematicus* (typhus with a spotted rash). Typhoid or enteric fever was recognized in the western world at the time of Hippocrates, by which early time it may have already been a widespread urban bacillary infection of the bowel. Louseborne *relapsing fever* probably evolved in the neighbourhood of the Middle East by adaptation of tick-borne spirochaetes to human lice. It seems to have been called 'relapsing fever' and distinguished from typhus during an epidemic in Edinburgh in 1843, described by Henderson. When epidemiological conditions are ripe for an epidemic of typhus, they are usually also ripe for one of relapsing fever. The two therefore often occur simultaneously, especially in wartime conditions. Epidemic louseborne *typhus* has similarly evolved from the adaptation to the human louse of the rickettsiae that are normally transmitted amongst rats by rat-fleas. It cannot be recognized in history with certainty until described in epidemic form in the fifteenth century by the Spaniards who named it *el tabardillo* or spotted fever (the name refers to a red cloak).

During these three centuries, typhus appeared in Europe and the British Isles intermittently in the form of devastating epidemics, especially associated with poverty and famine and with crowded conditions in jails and ships so that it also came to be known as jail fever and ship fever.[1] It was much feared as a disease that spread like wildfire and usually included some of the physicians, nurses, and others who attended the sick. The fever would begin suddenly and show no remission while the patient would rapidly become flushed and bloated, stuporose and delirious. If the victim were lucky, the fever would pass quite suddenly some two weeks later, leaving an extremely debilitated patient. In poorly nourished communities —and unhappily there were many of these—the mortality from this disease was high and at times rose to tens of thousands of deaths a month. When the fungus *Phytophthora infestans* almost completely destroyed the potato crop in Ireland in 1845–6 the resulting famine was associated with a disastrous epidemic of typhus. The Irish emigrated from their urban communities to Canada and the United States carrying their typhus with them. For example, records for 1847 indicate that, of 75 540 Irish emigrating to Canada, 30 265 contracted typhus and 20 305 died, 5293 of them dying at sea.

It was not until 1909 that Nicolle and his colleagues in North Africa showed that epidemic typhus was transmitted by lice. The organism responsible for Rocky Mountain spotted fever (a form of tick-typhus) was seen by Ricketts in the same year and that of epidemic typhus by several investigators in the next. Identities were confirmed by da Rocha Lima, who in 1916 called the organism of louse-typhus *Rickettsia prowazekii*, commemorating two foremost workers, H. T. Ricketts and von Prowazek, both of whom died of rickettsial infection in the course of their investigations.

Meanwhile, attention was being drawn in various parts of the world to distinctive diseases which were clearly typhus-like in nature but which occurred in quite different circumstances, namely as sporadic cases rather than as sweeping epidemics, in warm countries rather than in cold, in summer rather than

EMERGENCE OF THE TYPHUS GROUP OF FEVERS 65

in winter, and always associated with definite localities. This last feature has led to a fantastic number of place-names for these diseases, thus adding to the considerable confusion through which understanding of typhus fevers has passed. The relationship of these other fevers to epidemic typhus long eluded clarification, probably because the devastations of epidemic typhus and its dark history in Europe had given it a strong emotional hold on thought, leading for example to the perplexity of Major McKechnie of the Indian Medical Service when he found what appeared to be a typhus-like fever in apparently mystifying circumstances in the hills of northern India. 'After seeing her [the patient], I simply had to revise my notion,' he wrote in 1913 in an unpublished report, 'and then I found that the only thing against my thinking of typhus for other cases which had occurred was my obesssion as to the epidemiology. If some of the other cases were typhus then it must be my *obsession* that was wrong.'[2] (I shall have occasion to refer to this type of obsession later on.) That was before World War I. Between 1916 and 1922, there were over ten million cases of this disease in Europe and its influence on epidemiological ideas must still have been profound in the early '20's, but more and more evidence was collecting that there were, indeed, forms of typhus that occurred in particular places and did not spread from man to man as epidemics. The stage was therefore set for these endemic counterparts of typhus to be fully recognized, investigated, and set in their place whenever and wherever inspiration and opportunity happened to coincide.

By about 1910, the following six diseases were recognized but without the clear realization that they were all parts of the same jigsaw puzzle: 1. Epidemic typhus was well known; the others were endemic or place diseases which tended to be pinned down geographically in people's minds by being given place-names. Epidemic louse-borne typhus was already widespread and universally feared. 2. Rocky Mountain spotted fever was another severe infection, first observed in about 1873 among settlers in the Bitter Root Valley, Montana, and later in the Snake River Valley in Idaho.[3] All the typhus fevers are accompanied by macular rashes. In the case of the Rocky

Mountain infection minute haemorrhages frequently occur within the macules so that not only do they persist on compression, but also some staining remains afterwards—hence the name spotted fever. As early as 1902, Wilson and Chowning implicated the wood tick in transmission of Rocky Mountain spotted fever.[4] 3. Mossman fever, so named because it was first encountered in 1877 amongst the settlers on Daintree River at Mossman in tropical Queensland,[5] was also associated with valleys. 4. Japanese river fever or tsutsugamushi disease, described in accounts published in Japanese and in German between 1877 and 1879, was a severe infection characterized by the development of a small ulcer or eschar at the site of the infected bite by the vector mite.[6] 5. Brill's or Brill-Zinsser disease was for a long time the example of non-contagious fevers resembling typhus. Between 1898 and 1910, Brill collected a series of 200 perplexing cases which closely resembled louse-typhus but were milder and unrelated to lice.[7] We now know that most of them were in fact no more than relapses or recrudescences of louse-typhus in immigrants who had already recovered from louse-typhus acquired in Europe; but it can at once be seen how confusing these cases must have been to the earlier investigators until Zinsser clarified the problem in 1934.[8] 6. Finally, 'pseudotyphoid'[1] was identified in North Sumatra when 158 cases of fever broke out there in 1908 among labourers clearing a tobacco estate for replanting after it had lain fallow for many years. This estate had originally been cleared from forest but with no outbreak of fever. Schüffner and Wachsmuth suspected that the vector of pseudotyphoid was a mite.[10]

There is a curious coincidence about these many dates. This was a period of conquest of new territories, immigration and settlement by man, and these endemic diseases are well-known hazards of explorers and settlers. The timing in the case of the Brill-Zinsser episodes was largely due to recrudescences following a wave of immigrants from endemic countries. In the case

[1] These patients were investigated by Schüffner,[9] who was impressed by the similarity to typhus but even more so by the way in which the slower onset and disappearance of fever resembled the course of typhoid. He therefore called the disease *pseudotyphus abdominalis*. This should not be rendered as 'pseudotyphus' in English.

EMERGENCE OF THE TYPHUS GROUP OF FEVERS 67

of tsutsugamushi disease, the date was simply that of publication in the Western world stimulated by the emigration to Japan of Professor Baelz. A good account of the disease had already been published in 1810 by Hakuju Hashimoto.[11]

FIRST CLARIFICATION OF IDEAS: THE TYPHUS GROUP

Animals have in the past been awarded honours and even military decorations, while certain arthropods have been damned and accorded disrepute and one vector has been praised for the suffering it has caused a colonial 'enemy'.[1] But at least one other arthropod deserves commendation, and that is the tick that bit Major J. W. D. Megaw, I.M.S. (now Major-General Sir John Megaw) on the neck in 1916 while he was in the forests of the Himalayan foothills. Megaw developed a severe attack of what was clearly a typhus fever, and medical science ultimately gained by his misfortune. Megaw recognized the resemblance of his own fever to a Brill-Zinsser episode and later to a 'spotted fever' (Megaw 1917–21).[12] He applied himself with energy and inspiration to what had become a personal problem and very soon made a most important contribution by grouping typhus, Rocky Mountain spotted fever, tsutsugamushi disease, and the various related fevers of uncertain origin into a single typhus group of fevers caused by rickettsiae and transmitted by various arthropods. He also (in 1921) classified these typhus fevers according to their vectors. At that time there were large gaps and many uncertainties in this scheme, but we now know there could have been no more lasting or stimulating system of classification.

For at least the next two decades, it was generally believed that the organism of epidemic typhus, ever foremost in the minds of people, had somehow escaped into wild vectors and their animal hosts. That this view was taken is testimony to the powerful influence of tradition. It was indeed the second

[1] In Lagos a few years ago, one of the vernacular newspapers most vitriolic about former British Colonial 'oppressors' published a diatribe by a doctor in which he praised the mosquito for the curb it had placed on the white man's activities in West Africa. He suggested the mosquito should for this reason merit inclusion in the new flag of Nigeria. The harm done to West Africans by mosquitoes, tsetse, and other insects was glossed over.

such testimony in the history of typhus at that time. The first example was the hold on men's minds of the vivid clinical picture of louse-borne typhus and the fact that it occurred in devastating epidemics. As we have noted, this tended to hold people back from fully recognizing the existence of typhus fevers that did not fit into this pattern. We may remind ourselves of the history of leptospirosis, which was first identified in its most virulent form as Weil's disease, invariably accompanied by jaundice, frequently by haemorrhages, and usually by a high mortality. Both this clinical description and the epidemiological features were well advertised in textbooks, and the infection, contracted from rats' urine, was greatly feared. This picture delayed for many years the recognition of other forms of leptospirosis that were much milder and whose symptoms, such as aseptic meningitis without jaundice, were totally different. This obsession with a distinct clinical picture resembles the obsession of which McKechnie complained. It probably also helped to delay the proper recognition of 'pretibial fever', which occurred in North Carolina in 1940–1942, as a short fever accompanied by a rash over the shins. This was retrospectively diagnosed as a leptospirosis many years later by serological tests. Before that time, pretibial fever was attributed to an unknown virus. The second example of the influence of tradition on thinking was the deeply-rooted anthropocentric attitude that man was a unique creature suffering from diseases which were especially created for him. When confronted by typhus-like fevers contracted in the field, investigators automatically assumed that this was due to the transmission of human typhus to various arthropods and their animal hosts and to the maintenance and modification of the rickettsiae in these new circumstances. The mind rebelled against the biologically much more reasonable explanation that epidemic louse-typhus must have been a fairly recent human adaptation of an organism that had existed long before in different forms in various parasitic arthropods and their hosts.

Meanwhile, cases of the 'pseudotyphoid' of Schüffner continued to appear in estates in Sumatra until another big outbreak in 1922, this time among labourers who were clearing a neglected rubber plantation that had become overgrown with grass and secondary growth.[13] The 1922 outbreak was investigated by the physician to the Goodyear Tyre and Rubber Company, Dr E. W. Walch, and the estate medical officer, Dr N. C. Keukenschrijver, and their important findings were the basis for definitive research which was shortly to be carried out in Malaya. They emphasized the close resemblance of 'pseudotyphoid' to tsutsugamushi disease and discovered the vector, a mite which differed but slightly from the Japanese tsutsugamushi or akamushi and which Walch named *Trombicula deliensis* from the Deli area of the northeast coast near Medan where these outbreaks had occurred. This Sumatran mite-typhus differed from the tsutsugamushi disease of Japan by its milder course and by the occasional absence of an eschar, a difference to which we shall refer again.

Thus, by 1923, typhus fevers were known to comprise a very widespread group of diseases. The classical epidemic typhus, transmitted by human lice from one person to another, occurred in epidemics and was associated with lousiness. America, Africa and the Mediterranean region, and probably India had each developed its own form of typhus spread by ticks. A form in Japan and another in Sumatra were spread by chigger mites of the genus *Trombicula* (now *Leptotrombidium*). Finally, a large number of infections resembling epidemic typhus but generally milder and not occurring in epidemics were of unknown origin. A new attack on these uncertain fevers was next to be made in Malaya.

TYPHUS IN MALAYA

Between August 1924 and January 1925, 18 cases of fever in Malaya were diagnosed as typhus by Dr William Fletcher, the first cases being from the small mining village of Kepong near Kuala Lumpur. Fletcher arrived in Malaya in 1903, became Director of the Institute for Medical Research, the 'I.M.R.' in Kuala Lumpur, in 1926 and retired in 1927 when he was succeeded by Dr A. Neave Kingsbury.[1] In his last three years

[1] An account of the 'I.M.R. in K.L.', including a historical and cultural review of Western medicine in Malaya, was published in 1951 as a Jubilee volume: *The Institute for Medical Research, 1900-1950*, by various authors, Study No. 25 from the Institute for Medical Research, Federation of Malaya, Kuala Lumpur, Government Press, 389 pp.

he sifted leptospirosis, tropical typhus, and scrub-typhus from the confusion of undiagnosed fevers of Malaya.

The patients were given the usual meticulous clinical and laboratory examinations, including the Weil-Felix serological test for typhus, to which they reacted positively. Fletcher and Dr J. E. Lesslar, Assistant Pathologist, prefaced their account of these cases with the following words:

> An important discovery always opens the door to progress. The discovery of the Weil-Felix reaction has made it possible to recognise as typhus a continued fever of the Malay states which we have called Tropical Typhus. Moreover, we fully expect that the extended application of this test will show that the disease has a widespread incidence far beyond the confines of this small peninsula ... we call this variant 'Tropical Typhus' because it appears to be more common in the tropics than the epidemic form. It is necessary to distinguish it by some name; to call it simply 'typhus' is to mislead and alarm the public who, though they may be quite ignorant of everything else about typhus, know that it is highly infectious and may spread like wildfire.[14]

THE WEIL-FELIX REACTION: FURTHER CLARIFICATION

The application of the Weil-Felix test in Malaya involved one of those extraordinary accidents which occasionally stimulate scientific progress and enliven its history. So much has depended on this test that it is worth following its development in some detail.

In typhus, as in many other fevers, the damage to the small blood vessels of the bowel wall allows a variety of bowel organisms to gain entry to the blood and many of these are excreted in the urine. In the search for the causative agent of typhus, a number of organisms were recovered from the urine of typhus patients and these were tested to see if they could be agglutinated by the patient's serum. In 1909, an organism thus isolated from urine was found by Professor W. G. Wilson to be agglutinated by the serum of typhus patients. During an outbreak of louse-typhus in Galicia in 1915 Weil and Felix isolated an organism, the Proteus bacillus, from the urine of a patient. Suspensions of this were agglutinated by high dilutions

EMERGENCE OF THE TYPHUS GROUP OF FEVERS

of the serum from the patient and also from other cases of louse-typhus. In 1916, Felix isolated a strain of the Proteus bacillus that he called the 'X.19' (or OX.19) strain. It appeared to be sensitive and stable and it became the basis for the Weil-Felix reaction, widely used as a laboratory test for epidemic typhus. In cases of sporadic typhus, however, this test gave negative or feeble and generally unreliable results, although it was positive in Brill's disease and in an outbreak of sporadic cases in Australia investigated by Hone in 1923.

The sample of the strain which Fletcher and his colleagues used in Malaya had been taken from the British national collection of type cultures in 1921 and given to the Bland-Sutton Institute; from there it had been brought to Malaya in 1924 by Dr Kingsbury. The extraordinary accident to which we have referred is that this culture was, in fact, not an X.19 strain as was supposed but by some mischance or mutation was a new strain of Proteus bacillus; furthermore, by some happy miracle, it was agglutinable by the serum of scrub-typhus patients but not by sera from any other kind of typhus. There seems to be something more than coincidence in the arrival at this time of this particular new strain, for which there is no substitute.

This difference between the label on the bottle and its contents temporarily confused the issues even further.[15] The 18 Malayan patients studied in 1924 and early 1925 suffered from what we now know to have been tsutsugamushi disease and they gave positive reactions to Kingsbury—but this was supposed to be the X.19 strain. This at once appeared to link these cases up with Brill's disease. Fletcher and Lesslar wrote a preliminary account in a bulletin of the Institute for Medical Research, Kuala Lumpur, in 1925. They stated as follows:

Atypical Typhus—There is a sort of typhus which differs essentially from the ordinary type, by reason of its low infectivity; so far from passing easily and rapidly from the sick to almost every susceptible person who comes in contact with them, it is doubtful if the infection can pass direct from one person to another. Those suffering from this form of typhus can be nursed in the general wards of a hospital with impunity to the other patients, or they may remain at home without infecting their families, and it is unusual for more than one person to fall sick in the same household. The symptoms of this kind of typhus are identical with those of the classical form, and the blood gives a positive Weil-Felix reaction [based on use of the wrongly labelled bottle!], but the epidemiological features are so fundamentally different that they do not bear even a family resemblance to those of the great epidemic disease.[14]

The Weil-Felix reaction in Malaya soon gave some anomalous results, irreconcilable with those obtained elsewhere, and the next step in Malaya was to examine this tool more closely. After many laborious comparisons, the investigators unmasked the unique 'Kingsbury' or X.K. strain, agglutinated only by sera of scrub-typhus patients. The bottle could now be correctly labelled.[16, 17]

TWO KINDS OF TROPICAL TYPHUS: 'SHOP' AND 'SCRUB'

By applying this test, it was shown that 60 cases of 'tropical typhus' studied in 1926 comprised 32 X.K.-positive but X.19-negative, and 28 X.19-positive but X.K.-negative. The X.K. cases included two distinct outbreaks localized in particular rural areas. They involved cowherds and similar individuals who were exposed in the same way that tick-typhus patients were reported to have been exposed. In contrast, the X.19 cases were not localized; they included indoor workers, and at least seven of them occurred scattered about Kuala Lumpur itself. This at once suggested that the two forms had different vectors.

Meanwhile, cases of tsutsugamushi disease were occasionally recognized in Malaya. Dowden had reported a case in 1915 which appeared to be the same as Schüffner's cases of pseudotyphoid, which in turn were accepted as a mild variety of tsutsugamushi disease carried by a *Trombicula*. No more cases in Malaya were diagnosed as tsutsugamushi disease until four European planters were seen by Fletcher in consultation with Dr J. W. Field in 1926. Three of them came from an oil palm estate to the west of Kuala Lumpur, where they had been supervising clearing operations. They had typical eschars. Their Weil-Felix reactions were X.19-negative, which is as it should be; but, by one of the mischances so familiar to research workers, they were only very feebly and erratically positive to the X.K. strain. Had these cases been strongly X.K.-positive, as indeed they might readily have been (the reaction may develop late), then the relationship between them and the X.K.-type of tropical typhus would have been established very much earlier. As it is, two types of tropical typhus and the tsutsugamushi disease came to be considered as three distinct entities instead of only two. As late as 1930, in the Annual Report of the Institute, sections dealing with Tropical Typhus and Tsutsugamushi Disease are separated by 18 pages devoted to other topics. Futhermore, seeing cases side by side, the careful clinical observers had 'no hesitation in saying that they are distinct and separate diseases.'[18] Most of this confusion was due to tradition, to the voice of authority, to the 'obesssion' which McKechnie had confessed; this time, it was the emphatic statement by the Japanese authorities that the mortality of tsutsugamushi disease was very high and that the eschar, with its local lymphatic swelling or bubo, was invariably present— one is now tempted to add 'even when it cannot be seen!'. There are many such lessons to be learnt in the history of typhus fevers.

THE SECOND PHASE OF MALAYAN RESEARCH

The serological and epidemiological studies of Fletcher and Lesslar had by 1925 firmly established the existence of typhus-like fevers in Malaya. These were shortly to become as frequent as the enteric fevers. In 1926 an outbreak that first appeared in the oil palm estate to the west of Kuala Lumpur was followed by a second series of cases from an adjacent plantation in 1932 and for some years thereafter. Typhus research was then undertaken by four new workers at the Institute for Medical Research. One, Dr R. Lewthwaite, was appointed to the Institute as a Research Student in January 1927, and later as Pathologist. From 1931 onwards, in collaboration with Dr S. R. Savoor, Assistant Pathologist, he made the major contribution of Malayan research towards an understanding of the tropical forms of typhus. To this general picture B. A. R.

Gater, Entomologist, added a deeper understanding of the vectors.

Dr Ludwig Anigstein (1928–31), the fourth member of this group of workers, before he returned to the State Institute of Hygiene in Warsaw, devoted much effort to confirming the work of Fletcher, Lesslar, and Lewthwaite, and to studying strains of the Proteus and other bacilli; he was convinced that these bacilli were biological phases of the causal organism of tropical typhus. At that time it was not known that the rickettsiae of typhus are virus-like, are impossible to cultivate on artificial media, and demand living cells for growth. Here we have an example of an error in judgement leading an exceptionally able and energetic scientist into a large and fruitless investigation. This has happened to many scientists, whose abilities never gained recognition because they backed a losing horse.

During these years a painstaking search was made for a satisfactory laboratory animal. The unknown causative agent of tropical typhus could not be properly studied until it had been established in a suitable animal by a reliable technique. After the work of Mooser in 1928, the guinea-pig was regarded as the ideal experimental animal for epidemic louse-borne typhus and Rocky Mountain spotted fever, and nothing suggested that it would prove useless in Malaya. The Japanese had long since ceased to use this animal for work with tsutsugamushi disease, but the relationship between scrub-typhus and tsutsugamushi disease was not then realized. Lewthwaite and Savoor, however, ultimately established a strain in guinea-pigs whose resistance had been deliberately lowered by vitamin deficiency. Here another laboratory misfortune set these workers back; two out of three guinea-pigs had accidentally become infected with the spirillum of rat-bite fever, which, as was discovered later, gives a positive X.19 Weil-Felix reaction. Such freak accidents seem especially designed to rob scientists of their sleep or sanity. It is surprising that scientists have not learnt to tranquillize themselves by a belief in gremlins—a belief which, I am sure, has helped Air Force pilots and navigators to dissipate some of their exasperation harmlessly outward and to invoke their sense of humour rather than their resentment. It happens that the white mouse is the laboratory animal of choice for scrub-typhus rickettsiae, but this was not discovered until 1934, by Dinger in Sumatra.

Meanwhile, the mites were being studied energetically. Lewthwaite's pencilled marginal notes may still be found in Walch's publications on the Sumatran pseudotyphoid. Not only '*Trombicula*' *deliensis* but, apparently, *T. akamushi* was also identified. Gater took over Lewthwaite's collection in 1928. In the Annual Report of the Institute, the year 1928 is curiously signalized by the terse comment that a card-index filing system was introduced in the office. This must have been something of an augury; it was a year of solid achievements in several fields. Gater published a 'Part I' paper describing 14 new species of trombiculids. This set an exceptionally high new standard, which I regret was not then maintained by most workers on these mites; Gater's work remained exemplary for over two decades. Unfortunately, his later and would-be larger unpublished papers never appeared because his records and many drawings were destroyed in Singapore in 1941.

Lewthwaite and Savoor soon demonstrated that scrub-typhus, tsutsugamushi disease, the mite-typhus of Sumatra, and the coastal fever of North Queensland were one disease, varying locally in severity of the disease and incidence of the eschar. The urban 'shop-typhus' was identified with an apparently widespread form of typhus transmitted by rat-fleas.[19] They urged that the names 'rural typhus' and 'scrub-typhus' be dropped in favour of tsutsugamushi disease, but wartime usage firmly established the simple expression 'scrub-typhus'.

Using Megaw's classification, however, the disease would be referred to as 'mite-typhus', in contradistinction to louse-, flea-, and tick-typhus. Here, however, this system of classification breaks down in a way that is interesting to those who respect the meanings of words. The subclass Acarina is known by two popular English words: *ticks*, referring to a small highly specialized group of wholly parasitic large acarines, and *mites*, referring to all the remainder. This is counter to Nature's way of dividing the Acarina, which is into two other groups with different evolutionary origins. Some workers regard these last two groups as two subclasses. The first one, fairly homogeneous, includes the ticks (Metastigmata) and also many related mesostigmatic mites (sometimes known popularly as parasitoid or gamasoid mites). The other unrelated groups, very large and much more heterogeneous, include the beetle-mites, scabies-mites, and feather-mites and a large order (Prostigmata) in which the larvae are often parasitic. In this last group are the trombiculid vectors of scrub-typhus (see Appendix 1).

In 1946 and the following years, an apparently new form of typhus called rickettsialpox[20] was recognized in Boston, New York, and neighbouring areas. It is transmitted by a species of mesostigmatic mite infesting mice. These mites are very much more closely related to ticks than they are to trombiculids. It appears that rickettsialpox is caused by a spotted-fever-like tick-borne rickettsia that has become adapted to something popularly called a 'mite'. Many workers have grouped rickettsialpox with mite-typhus (scrub-typhus) instead of where it belongs, with tick-typhus. This sort of confusion could never have arisen among the Russians who use the one word *'kleschchei'* for all the Acarina, using adjectives to describe a particular type, e.g. *kleschei krasnotelye* or red-*kleshchei* or simply *krasnotelki*, the red-bodied ones, for trombiculids.

EVOLUTION OF THE TYPHUS GROUP OF FEVERS[1]

It is now possible to speculate profitably on the interrelationships between the typhus fevers. The main elements that concern us are the causative rickettsia, the vector, the vertebrate maintaining host (whether it is man or animal), and their world distribution.

The rickettsiae belong to a large and somewhat heterogeneous group of highly modified and specialized bacteria that are somewhat virus-like, usually requiring living cells for their multiplication. It appears that the rickettsiae have become virus-like by starting off as bacteria and becoming increasingly specialized by their adaptation to parasitic life within the cells

---

[1] The arguments presented in this section are dealt with in more detail in a paper in MS by Audy and Marchette,[21] to be published, and also to some extent by Audy.[22]

EMERGENCE OF THE TYPHUS GROUP OF FEVERS

of arthropods, until they have become wholly dependent on living cells. Some groups of rickettsia-like organisms have evolved further in their arthropod hosts by developing a form of interdependence known as mutualism, in some cases producing vitamin-like substances which their hosts require for survival: e.g. the sheep-ked and the common bedbug have permanently indwelling rickettsial organisms, the ked-rickettsiae multiplying within the gut cavity and the bug-rickettsiae multiplying in a special organ (the mycetome) reserved for the purpose. Other groups of rickettsiae have evolved further by becoming adapted to the vertebrate hosts on which their invertebrate arthropod hosts feed parasitically. The typhus-group of rickettsiae is an example. Yet other groups of rickettsiae have progressed even further by becoming independent of their original arthropod vectors and being transferred from vertebrate to vertebrate. One may see this happening with the rickettsia of Q fever, an aberrant member of the typhus group. This rickettsia, *Coxiella burneti*, is in nature transmitted normally by ticks among their animal hosts, but the organism is passed in huge numbers in milk and the birth-fluids so that transmission amongst livestock and people handling them has become independent of arthropod vectors, at least temporarily and locally. These latter groups of rickettsia-like organisms have *possibly* been joined by other bacteria that have gained their rickettsia-like character by developing obligatory intracellular parasitism of vertebrate cells without the intermediate agency of an arthropod—but we have no real evidence to support this idea. Other rickettsia-like organisms, whose origin is wholly unknown, are responsible for conjunctival and venereal infections or both (trachoma, inclusion conjunctivitis, lymphogranuloma venereum, abortion and arthritis in animals, etc.) and a related group for pulmonary infections derived from birds (psittacosis, ornithosis).

All that follows applies only to the related rickettsiae that form the 'typhus group'. Speculations about the phylogeny of this group may have no bearing whatsoever on the other rickettsiae or rickettsia-like organisms.

The main features of rickettsia, vector, host and distribution are conveniently tabulated in Table 1. Classical louse-typhus

TABLE 1. The Typhus Group of Fevers: Rickettsiae, Vectors, and Distribution

	Disease	Vector	Rickettsia	Distribution
A.	*Classical Typhus Group*			
	Louse-Typhus	Human lice	*R. prowazekii**	Worldwide urban—coincident with body-lice but increasingly under control. Major centre in Europe.†
	Flea-Typhus	Rat-fleas	*R. typhi* (*R. mooseri*)	Worldwide urban—coincident with house-rats.
B.	*Tick-Typhus Group*			
	Tick-Typhuses	Ticks	*R. rickettsi* & related forms	Local in all continents.
	Rickettsialpox	Mesostigmatic mites	*R. akari*	Eastern USA, NW Asia.
C.	*Mite-Typhus Group*			
	Scrub-Typhus	Trombiculid mites	*R. tsutsugamushi* (*R. orientalis*)	Restricted—Japan-India-Queensland.
D.	*Aberrant Forms*			
	Trench Fever	Human lice	*R. quintana*	Europe, America, where lice prevail.
	Q Fever	Ticks; also directly transmitted from livestock	*Coxiella burneti*	Worldwide.

* According to the rules of botanical and zoological nomenclature, a person's name is first latinized, to agree in gender with the genus, and then placed in the genitive case to complete the name of the species. Thus Diard through Diardius yields *Rattus diardii*. This is the origin of the double-*i*'s, as in *prowazekii*. Botanists, bacteriologists, and now virologists have slightly different rules, but many do not observe them all. Recent modernization of rules has led to a return to single *i*'s except for *prowazekii* which was originally spelled that way.

† Ten years ago Reiss-Gutfreund[23] reported evidence of infection of livestock and their ticks by *R. prowazekii* in Ethiopia. Philip[24] has recently reviewed work in Ethiopia, Egypt, and Latin America. We may assume that the rickettsia,

and trench fever are purely human infections, although the causative agent of the former has been reported from domestic animals and their ticks. All the others are infections maintained primarily among rodents (or the related rabbits).

In group A and in trench fever (group D), the vectors are *insects*, to which the rickettsiae are restrictively or incompletely adapted. These rickettsiae are confined to the cavity of the gut and infection is by the introduction of infected faeces. *R. prowazekii* and *R. typhi* are usually fatal to human lice that ingest them.

All the other rickettsiae are transmitted by *acarines*, to which they are very intimately adapted (except perhaps for *R. akari*), passing through the gut-wall to be dispersed in the tissues. They are thus injected directly with the saliva and even infect the eggs so that infection is passed to a proportion of the successive generations of larvae (transovarial transmission).

In brief, the evidence suggests that the acarine forms of rickettsiae are the most primitive. Either a group B tick-borne or a group C chigger-borne rickettsia became adapted through the rodent host to bloodsucking insects such as rat-fleas and rat-lice. Man, living in rat-infested dwellings, often became infected by rat-fleas (and also by rat's urine); and the circulating rickettsiae, repeatedly ingested by human lice, readily became adapted to these. This changed the whole character of the infection, from endemic (picked up from time to time from a particular place) to epidemic (spread from man to man as a truly human disease). Thus arose *R. prowazekii* from *R. typhi*, and *R. typhi* in its turn from group B or C rickettsiae.

There appear to have been at least two further steps in the evolution of the insect-borne rickettsiae, both consequences of further adaptation between the rickettsiae and their insect and human hosts. One led to the emergence of *R. quintana*, fully adapted to the louse and approaching full adaptation to man, for it produces mild relapsing attacks. The other led to the emergence of one or perhaps more species such as *R. rochalimae*, which do not affect the louse and cause no harm to man (therefore not listed in Table 1). In other words, a rickettsial species-complex has evolved in the louse and man.

Group A rickettsiae are cosmopolitan in distribution because they have been spread by man, either directly by the movement of louse-infested people, or indirectly through the part man has played in encouraging the spread of the two house-rats, *Rattus rattus* and *Rattus norvegicus*, round the world from their original home in Asia. The flea followed the rat. The chief flea concerned is our old acquaintance, the plague-flea.

The primitive rickettsiae in ticks and mites comprise two totally different groups in two very different types of acarines. *Rickettsia akari* (rickettsialpox) seems to be an adaptation of the *R. rickettsi* complex (tick-typhus) to a mesostigmatic mite, *Allodermanyssus sanguineus*, parasitic on house-mice. Also, at least two well-recognized species of tick-borne rickettsia (*R. dermacentrophilus* and the organism of maculatum-disease of guinea-pigs) appear to have budded off the parent *R. rickettsi* stem and are non-pathogenic to man.

Ticks and trombiculid mites are phylogenetically so far removed from each other (see page 154) that we must, in the interests of economy of hypothesis, choose one or the other for the parent stem. At first sight, the ticks seem the more likely choice because forms of tick-typhus are worldwide in distribution, while scrub-typhus is geographically localized. But then we must ask how the ticks got the rickettsia in the first place. It does not help to pretend they obtained these from the vertebrate host, because it is the arthropod which is the primitive and primary host of these rickettsiae.

In brief, I suspect that the ultimate origin of the rickettsiae that occur as highly specialized intracellular parasites of arthropods is that great bacteriological laboratory, the soil, which may contain a ton or more of micro-organisms per acre. The proposed evolutionary sequence would have been: (a) repeated ingestion of certain soil-bacteria by vegetarian or scavenging arthropods, and adaptation of these to parasitic life as rickettsioid bodies within these arthropods or their cells; (b) further adaptation of these rickettsioid organisms to intracellular parasitic life in predatory arthropods which feed on the vegetarian arthropods; (c) adaptation to a new kind of host, the vertebrate, during the evolution of vertebrate parasitism by the arthropod (since the step from predation to parasitism on a larger creature is short and has been taken by several different groups of arthropods). Thus we have a rickettsial organism being maintained by a parasitic arthropod and its vertebrate host, possibly causing disease at first until there is mutual adaptation (cf. the adaptations beyond *R. prowazekii* into benign rickettsiae).

Now we can state our problem as a simple question—which of the two, tick or trombiculid, has in its phylogenetic past had opportunities to pick up rickettsial organisms by preying on soil predators or on soil arthropods generally? The ticks are a highly specialized and wholly parasitic group. To trace them back to a parent stem with the mesostigmatic mites leads to complexities which are discussed elsewhere.[21] The trombiculids, in contrast, as well as the trombidiid stem from which they evolved, are parasitic only in their larval stages while the post-larval stages are predatory on soil arthropods. Economy of hypothesis seems to demand the conclusion that the rickettsiae concerning us may probably be traced back to a primitive stem in trombiculid mites. Such rickettsiae, circulating in rodents, have become successfully adapted to ticks feeding on the same rodents, followed by the emergence of a distinct rickettsia that has undergone further adaptations to various local tick-host combinations, leading to geographical and ecological subspeciation (the geographical forms of tick-typhus).

This argument has been presented with a view to simplicity. It is not in the logical sequence in which it developed. The logical sequence adopted was to follow the ticks and the trombiculids backwards in time to see how either could possibly have acquired rickettsiae. This led back to the soil.

Why should scrub-typhus be geographically localized and tick-typhus widespread if, as hypothesis suggests, the parent stem of the typhus group of rickettsiae was in the trombiculids? It may be that this is because a primitive and relatively silent trombiculid rickettsiosis in sundry mammal-chiggers in Southeast and East Asia have been involved in a major biological explosion: the evolution of *Rattus*. Southeast Asia is the centre of evolution and dispersal of the genus *Rattus*, of which there are over 500 named forms in the Malaysian subregion alone. [25, 26] We may infer that this silent primitive rickettsiosis in mammal-chiggers may be worldwide but no longer generally infectious to mammals except for the special developments around Southeast Asia; transfer of infection to ticks must then have been far back in time. There are several alternatives but this is not the place to toy with them.

Man has greatly encouraged the development of the rat. By the attractiveness of his own houses and fields, both cultivated and abandoned, he has led to their exploitation by at least some of the enterprising forms of rat. He has cleared and converted vast areas of forest to grasslands and bamboo tracts by shifting cultivation (a method known as slash-and-burn cultivation). Various local rats, adapted to living in fields and houses, have emerged in the new conditions. They are indeed animal weeds among the plant weeds. Two of the most successful species, *Rattus rattus*, the climbing-rat that favours the roof, and *R. norvegicus*, the ground-rat that prefers the drains, have spread round the world with man's help. (So also has the house-mouse, *Mus musculus*, from this same general region of Asia.) We may conjecture that a silent forest rickettsiosis—whose supposed existence is supported by evidence—has been brought out into the open and down to the ground by adaptation to the combination of two animal weeds, namely, the populous field-rats and a few species of *Leptotrombidium* adapted to these conditions. The multiplication and spread of the rickettsia has then been tremendously boosted in conditions where 10–50 or more larvae of *L. akamushi* or *L. deliense* may feed on a single rat every day.

I shall refer again to aspects of evolution in the two following lectures.

REFERENCES

1 ZINSSER, H. (1934). *Rats, Lice, and History. Being a Study in Biography, which after Twelve Preliminary Chapters Indispensable for the Preparation of the Lay Reader, Deals with the Life History of Typhus Fever.* Boston, Little Brown, p. 32.
2 MCKECHNIE, D. (1913). Unpublished report, quoted by J. W. D. MEGAW. [12]
3 MAXEY, E. E. (1899). Some observations on the so-called spotted fever of Idaho. *Med. Sentinel, Portland, Oregon* 7, 433–8.

4 WILSON, L. B. and CHOWNING, W. M. (1902). The so-called 'spotted fever' of the Rocky Mountains. A preliminary report to the Montana State Board of Health. *J. Amer. med. Ass.* **39**, 131–6.
5 BREINL, A., PRIESTLY, H. and FIELDING, J. W. (1914). On the occurrence and pathology of endemic glandular fever, a specific fever occurring in the Mossman district of North Queensland. *Med. J. Aust.* **1**, 391–5.
6 BAELZ, E. and KAWAKAMI, S. (1879). Das japanische Fluss- oder Überschwemmungsfieber, eine acute Infectionskrankheit. *Virchow's Archiv. path. Anat.* **78**, 528–30.
7 BRILL, N. E. (1910). An acute infectious disease of unknown origin. A clinical study based on 221 cases. *Amer. J. med. Sci.* **139**, 484–502.
8 ZINSSER, H. (1934). Varieties of typhus virus and the epidemiology of the American form of European typhus fever (Brill's disease). *Amer. J. Hyg.* **20**, 513–32.
9 SCHÜFFNER, W. (1915). Pseudotyphoid fever in Deli, Sumatra (a variety of Japanese Kedani fever). *Philipp. J. Sci.* **10** (Sec. B), 345–53.
10 SCHÜFFNER, W. and WACHSMUTH, M. (1910). Über eine Typhusartige Erkrankung (Pseudo typhus von Deli). *Z. Klin. Med.* **71**, 133–56.
11 HASHIMOTO, H. (1810). [Quoted by R. KAWAMURA (1926). Studies on tsutsugamushi disease (Japanese flood fever). *Univ. Cincinnati med. Bull.* **4** (Spec. Nos. 1, 2).]
12 MEGAW, J. W. D. (1921). A typhus-like fever in India, possibly transmitted by a tick. *Indian med. Gaz.* **56**, 361–71.
13 WALCH, E. W. and KEUKENSCHRIJVER, N. C. (1924). On pseudo typhus of Sumatra II. Some notes on epidemiology. *Trans. Far-East Ass. trop. Med. 1923*, 627–43.
14 FLETCHER, W. and LESSLAR, J. E. (1925). Tropical typhus in the Federated Malay States. *Bull. Inst. med. Res. F. M.S.*[1] No. 2 of 1925.
15 FLETCHER, W. and LESSLAR, J. E. (1926). Tropical typhus and Brill's disease. *J. trop. Med. Hyg.* **29**, 374–8.
16 FLETCHER, W. and LESSLAR, J. E. (1925). A comparison of some strains of B. Proteus employed in the Weil-Felix reaction. *Trans. Far-East. Ass. trop. Med. 1925*, **2**, 775–86.
17 FLETCHER, W. and LESSLAR, J. E. (1926). The Weil-Felix reaction in sporadic tropical typhus. *Bull. Inst. med. Res. F.M.S.* No. 1 of 1926.
18 FLETCHER, W. and FIELD, J. W. (1927). The tsutsugamushi disease in the Federated Malay States. *Ibid.* No. 1 of 1927.
19 LEWTHWAITE, R. and SAVOOR, S. R. (1936). The typhus group of diseases in Malaya. I. The study of the virus of rural typhus in laboratory animals. *Brit. J. exp. Pathol.* **17**, 1–14; II. The study of the virus of tsutsugamushi disease in laboratory animals. *Ibid.* **17**, 15–22; III. The study of the virus of the urban typhus in laboratory animals.

[1] Since World War I, the '*F.M.S.*' (Federated Malay States) in the journal references to Bulletins and Studies of the Institute has been replaced by '*Malaya*' (but not in the *World List*).

Ibid. **17**, 23–34; IV. The isolation of two strains of tropical typhus from wild rats. *Ibid.* **17**, 208–14; V. The Weil-Felix reaction in laboratory animals. *Ibid.* **17**, 214–28; VI. The search for carriers. *Ibid.* **17**, 309–17; VII. The relation of rural typhus to the tsutsugamushi disease (with special reference to cross-immunity tests). *Ibid.* **17**, 448–60; VIII. The relation of the tsutsugamushi disease (including rural typhus) to urban typhus. *Ibid.* **17**, 461–5; IX. The relation of tsutsugamushi disease (including rural typhus) and urban typhus to Rocky Mountain spotted fever (with special reference to cross-immunity tests). *Ibid.* **17**, 466–72.

20 HUEBNER, R. J., STAMPS, P. and ARMSTRONG, C. (1946). Rickettsialpox—a newly recognized rickettsial disease. I. Isolation of the etiological agent. *Publ. Hlth. Rep.* **61**, 1605–14.

21 AUDY, J. R. and MARCHETTE, N. (1966). The typhus group of fevers: evolution and relationships of the rickettsiae. In MS, to be published.

22 AUDY, J. R. (1959). The epidemiology of scrub typhus. *Proc. 6th Intl. Congr. trop. Med. Malar., Lisbon 1959*, **5**, 625–30.

23 REISS-GUTFREUND, R. J. (1956). Un nouveau reservoir de virus pour *Rickettsia prowazeki*: Les animaux domestiques et leurs tiques. *Bull. Soc. Path. exot.* **49**, 946–1021. (See also: *Rickettsia prowazeki* in domestic animals and their ticks. Presented at First International Congress of Parasitology, Rome, 1 September 1964.)

24 PHILIP, C. B. (1965). Epidemic typhus: preliminary evidence of rickettsial zoonoses in Latin America. In Document RES 4/2A, 26 April 1965, 'Research Activities of PAHO in selected fields, 1964–65'. Advisory Committee on Medical Research, Pan American Health Organization, Pan American Sanitary Bureau, Regional Office of the World Health Organization, Washington, D.C.

25 HARRISON, J. L. and AUDY, J. R. (1951). Hosts of the mite vectors of scrub typhus. I & II. *Ann. trop. Med. Parasitol.* **45**, 171–94.

26 HARRISON, J. L. (1949). The domestic rats of Malaya. *Med. J. Malaya*, **4**, 96–105.

4

THE IMPHAL CIRCUS[1]

THIS will be a personal lecture concerned with things as they appeared to me during and immediately after the war. From early 1945 until April 1946, a field laboratory of the British Army was working on scrub-typhus on the outskirts of Imphal near the Indo-Burma border. The development of this *ad hoc* laboratory had been made my responsibility. In October 1945 when I visited South East Asia Command Headquarters in Singapore, I learned accidentally that our Imphal laboratory out there in the wild was known in Headquarters as 'Audy's Circus', probably somewhat ruefully, but, I like to think, at least with some amiable toleration. I would be remiss in these lectures if I were to neglect some of the lessons I learned whilst acting now as ringleader, now as circus clown, at Imphal. But although this is a personal lecture I prefer the less personal title of 'The Imphal Circus'.

The Himalayan mountains branch southwards at the level of Burma, narrowing down to the knife-like ridges of the Arakan Yomas, parallel with the Indian Ocean coastline. Near the southwest tip of Burma, these mountains dip into the sea and continue southwards and then eastwards in a great sweep, emerging from the ocean at intervals to form the Andamans, the Nicobars, and the islands of Indonesia. Apart from a number of wild jungle paths, there are three main routes across these formidable mountains between India and Burma, all three of them tortuous and difficult. The Ledo Pass connects Assam with the rain-forests in the far north of Burma. Another route goes southwards down the Indian Ocean coastline and through the southern part of the Arakan Yomas into the rice-fields of lower Burma. Between these, the third route runs through a broad part of the mountain range where it is

[1] The work of the 'Imphal Circus' or the British Army Scrub-Typhus Research Laboratory, South East Asia Command, at Imphal, Manipur, 1945–6, has been summarized by Audy and Harrison[1] and in the abstract of a War Office report.[2]

separated from the plains of north Burma by the Chindwin River. This approach is from the Brahmaputra River through the tiny station of Dimapur (Manipur Road) up into the high mountains of Kohima, then southwards for 60 miles to the Imphal plain, a rice-bowl 40 miles long by 20 miles wide at an altitude of 2600 feet.[3] This is the centre of Manipur State, where the national sport is hockey on horseback, a game originating in Persia which has spread around the world as polo. (The world's first polo club was at Silchar near Imphal.) Northwards, towards Kohima, live the Nagas, those magnificent warriors whose huts, at the time we are talking about, were still adorned with grisly relics of head-hunting days. In the hills around Imphal were the Kukis and farther south, the Chins.

There are two routes out of the Imphal plain into Burma. One follows the Manipur River southwards past Tiddim and ultimately eastwards to join the Chindwin. A huge landslide in the Pleistocene period blocked the Manipur River and created the lake that slowly silted up to form the Imphal plain. The other route leaves the southeast corner of the Imphal plain through Palel and over 30 miles of wild and magnificent mountains to descend to the Kabaw Valley which runs parallel with the Chindwin. One may then either go eastwards to cross a mountain range to the Chindwin or track southwards along the notoriously unhealthy Kabaw Valley until one emerges on the Chindwin at Kalewa. In the southern part of the Imphal plain, a route leads westwards from Bishenpur across the mountains to India. During 1942 some 190000 refugees from Burma poured through Imphal. It was in March 1944, that, after creating a diversion in the Arakan, the Japanese 15th Army under General Mataguchi made its powerful thrust from three bases along the Chindwin until both Imphal and Kohima were surrounded and the road between them was cut off. The tremendous importance of the sieges of Kohima and Imphal was hardly realized by the outside world at the time. The loss of these remote places would probably have meant the loss of much of India. At Kohima, a vigorous battle was to rage for three days between opposing units dug in on both sides of the district officer's tennis court. Shortly

THE IMPHAL CIRCUS

after, it was through Imphal that the main advance of the Allied 14th Army was made into Burma.

Before the main Burma campaign, in which British-Indian forces suffered some 5400 cases of scrub-typhus, there had been several hints of things to come. For example, in September 1941 there was an outbreak of 107 cases of scrub-typhus in a unit in central Burma (Meiktila, south of Mandalay). From June 1942 through the rainy season there were 20 cases in Calcutta to be followed by 130 more the next year. These infections were acquired in waste land and park land in Calcutta, but it took people by surprise to find 'rural' scrub-typhus occurring within a town. We were later to find cases in Mandalay, Rangoon, Kuala Lumpur, and other towns. In early 1943 there were 33 cases in troops doing jungle training at Ranchi in India, and in the autumn of 1943 there occurred 58 cases among people camped among the rice-fields at Jhingergacha near Calcutta.

At the end of 1943, there occurred two outbreaks that seriously alarmed the military command. In October two companies patrolling a ridge on the approach to Tamu in the Kabaw Valley reported 121 cases of scrub-typhus. This ridge was later known as Mite Hill. At the time these cases were appearing, the 11th East African Division was engaged in large-scale jungle training exercises in south Ceylon. One unit of less than brigade strength carried out a four-day operation near the hamlet of Embilipitiya. The result was an outbreak of no less than 756 cases of scrub-typhus.[1] At that time I was a malariologist to the East African Division. Immediately after the outbreak, on my way to a survey of the southeastern part of Ceylon, I was able to inspect this focus and trap a few rats in the scrub. They were infested by the vector mite, *Leptotrombidium deliense*, at that time known to be a vector of scrub-typhus in Malaya and Sumatra but hitherto unsuspected in Ceylon. During the next two weeks we were engaged in field work in lonely places, and I had time to think about some of the striking impressions I had gained while camping in the Embilipitiya focus.

[1] This explosive outbreak is only one of several kinds that are described in Appendix 8.

PECULIARITIES OF THE CEYLON OUTBREAK

At the time of this outbreak, scrub-typhus was almost unknown in Ceylon. A number of cases were however diagnosed retrospectively once the authorities were alerted. At the time I went to Embilipitiya, scrub-typhus was regarded as a rare possibility and this influenced my thinking considerably. There had been many large military manoeuvres without demonstrating infected areas. Many questions were raised but they could be summed up in two words: Why Embilipitiya?

The Embilipitiya focus was an island of infection, a true 'typhus island'—I had not then heard of the term *yudokuchi* or 'poisonous place' used by the Japanese to describe such foci. The classical Japanese *yudokuchi* were grass-covered banks or islands along the course of certain rivers in northwest Japan. The Embilipitiya *yudokuchi* was a patchy area of waste land and fallow fields between two rivers, overgrown with rank *illuk* grass and other weeds. Adjoining it were cultivated fields. The patches of overgrown waste land in which the infections seemed to have taken place were unexceptional and similar to patches of waste land that could be found elsewhere in Ceylon.

The natural vegetation of Ceylon is forest, but the Embilipitiya *yudokuchi* was in a deforested area subjected to *chena* or slash-and-burn cultivation. To an epidemiologist, this provided a reason for the presence of this particular focus. The encouragement of the vector mites in large numbers depends upon the presence of numerous field rodents. Conditions seemed ideal at Embilipitiya because overgrown grassy areas were mixed with fields cultivated for millet, thus providing shelter for rats and adjacent to a plentiful food supply; for much millet gets wasted and becomes available to rats. At that time the literature which I had hurriedly read described several conditions in which the infection had appeared.[1] The natural vegetation of the endemic areas in Japan is a mixed deciduous forest, but the infectious areas were strictly localized in grassy islands and banks of fertile silt. In Sumatra, the Senembah Company cleared the native forest with no casualties. Tobacco was

---

[1] These have been described in the preceding lecture.

planted and then the plantation was allowed to run fallow for several years. An outbreak of scrub-typhus occurred among workers clearing the grassy overgrown plantation in 1908. In the case of the Goodyear rubber plantation in Sumatra, forest had again been extensively cleared with no casualties from scrub-typhus; but an outbreak[1] had occurred among workers clearing grass from an overgrown part of the plantation abandoned during the slump in the price of rubber. In both these Sumatran examples, the native forest seemed to have been harmless but infected areas of scrub-typhus had actually been created by artificial and man-made conditions. Similarly the main outbreak in the oil palm estate in Malaya in 1926 was among people who were clearing the tall grass away from the bases of young oil-palms that had been left to grow unattended for three or four years. There was no record of infections in workers when the forest was originally cleared for planting. In Ceylon, innumerable exercises in forest had yielded no infections.

In Formosa, however, Hatori had described scrub-typhus as occurring in a great variety of conditions including forests, for example, among camphor collectors. Nevertheless, there was no clear evidence that the infections had been incurred in the forest although the collectors had worked there—the *yudokuchi* themselves had not been identified in forest, and the infections might have been from isolated abandoned clearings.

Meanwhile, reports had been coming in of scrub-typhus in other war theatres. It was frequently associated with patches of *kunai* grass in New Guinea and Queensland and of *kogon* grass in the Philippines. In Malaya and Sumatra the grass was *lalang*. I found that *illuk*, *lalang*, *kogon*, and *kunai* were all local names for the same species of rank grass, *Imperata cylindrica*, a notorious, extremely widespread and vigorous weed. This coarse and rasping grass, unpleasant to travel through, possesses

[1] Scrub-typhus may occur in epidemic proportions when large bodies of people are exposed to risk together, but it is an endemic and not an epidemic disease; hence the use of the term *outbreak*, which editors of a few journals delete in the belief that it is a synonym of 'epidemic'. An outbreak of scrub-typhus does not develop and die down in the same way as an epidemic of a disease such as louse-borne typhus or relapsing fever.

rhizomes which make it fire-resistant, and it puts out plentiful supplies of fluffy seeds which are readily distributed by the lightest air currents. Local inhabitants cannot resist firing the coarse grass during the dry season. The grass thrives, but all seedlings are destroyed except for those of a few fire-resistant kinds such as some species of oak. Once this vigorous weed takes over a forest clearing, the tribesmen are forced to abandon it and clear another patch of forest for their cultivation. This grass, together with other weeds and the rapid loss in fertility of the soil on clearings in the tropics, encourages the primitive practice of shifting cultivation, also known as slash-and-burn cultivation.[4] This practice soon leads to the replacement of forest by a patchwork of grassland and scrub in various stages of regeneration. Later, vast tracts of forest become converted into extensive grasslands with narrow relict belts of gallery forest following the steep-sided watercourses. In these conditions, the annual grass fires that preserve this type of vegetation make a beautiful sight, glowing garlands festooning the hills night after night. It is a common practice in many places to set fires in such a way as to drive game to places where hunters lie in wait.

While I was in Ceylon I became very excited about *Imperata* and similar grasses and, indeed, about weeds in general for two reasons. One concerned the character of the hyperendemic foci already described in the literature and then being reported from theatres of World War I. The other was connected with the possibility that the post-larval stages of the vector chiggers might feed on plant juices. Japanese workers had already reported this—or rather they had been misquoted when they voiced the suspicion that the mites *seem* to feed on the juices of grasses and other weeds.[5] Mehta in India had reared *L. deliense* by 'feeding' the post-larval stages on slices of potato.[6] For a time, many investigators, including myself, were deceived by these accounts. We overlooked the statement of Oudemans in 1912[7] that the post-larval stages were predatory, feeding on other soil arthropods. It was not until 1945 that our eyes were opened when Wharton in Guam[8] and Jayewickreme in Ceylon[9] independently rediscovered the truth and reared nymphs and adults by feeding them on mosquito eggs. Mehta's survivors were cannibals, having fed on other nymphs and not on the potato juice. This discovery was too late to help us with our wartime struggles with these mites. Meanwhile, I misled myself in Ceylon into wondering if the dense mat of surface rootlets in some types of grassland might not favour sap-sucking mites. If so, it would help to account for the apparent association of scrub-typhus with grassy patches which also provided excellent living and foraging conditions for small omnivorous rodents and thus might encourage chiggers in both the parasitic and non-parasitic stages.

Three concepts grew in my mind from a comparison of the Embilipitiya *yudokuchi* with the others described in the literature. These concepts were of 'animal weeds', 'jungle tsutsugamushi', and 'man-made maladies'.

THE CONCEPT OF ANIMAL WEEDS

One striking feature of a community of plants and animals is usually unnoticed because it is taken for granted: this feature is the reliability of the community's composition. Decade after decade one may return to a woodland to find the same plants, insects, small mammals, and birds in roughly the same numbers behaving in the same way daily, nocturnally, and seasonally. Almost every species has the reproductive potential to overrun the forest, but populations are kept in check by external conditions such as food supply and internal mechanisms such as their neurosecretory systems. Even fluctuating populations are kept in some sort of check. This is the balance of nature. It is present because the entire community has evolved as a unit through geological time. Each different species has been genetically guided in its evolution by the other species that affect it. The result is a system with built-in feedback checks. The system may be as simple as that in a desert or as enormously complex as that in a tropical rain-forest. The nature of the system, known as an ecosystem, and its complexity are decided primarily by the climate but also by the terrain and the underlying rocks and soil. Any given set of conditions can permit a certain variety, complexity, and density of living forms. Given a set of conditions of climate, terrain, and soil, the richest assemblage it will permit is known as the local *climax*.

If climax forest is cleared, it becomes colonized by various short-lived and hardy herbaceous plants and starts on a slow development towards a life-form and composition that is very close to the original forest ecosystem. Such a developing assemblage is a *sere*, serially unfolding. The next year it will be more complex. One day the patch of waste land will become forest again.

The vigorous annuals of the early pioneer community of vegetation are followed by perennials and later by woody shrubs and trees. Some species of plants are characteristic opportunists, running riot in the pioneer communities. They are frequently not native to the local climax but are invaders from some other and more simple climax where conditions have preadapted them to development in the early disturbed conditions that follow destruction of climax vegetation. Man knows them in his fields and gardens as *weeds*. In their native climax vegetation such weeds are well-behaved members of their communities but in pioneer communities they have few restraints.

In its native habitat in Africa, the coarse grass *Imperata cylindrica* is a natural component of savannah. Its fluffy white seeds have earned it local names such as *msufi wa bara* or roadside-cotton, drawing attention to a habitat that indicates its potentialities as a weed. From its original home it has spread around the world, often with man's aid. Many efforts have been made to eradicate this grass, but it is too vigorous and its fire-resistant properties, combined with the behaviour of man, allow it to resist most natural competition. Vain attempts have been made to use the grass, but it is too coarse for fodder, not good for brewing beer, and is uneconomical to harvest for making paper. Fortunately, it cannot resist shade; therefore an overgrown clearing will gradually give way to scrub if it is protected from fire. Such protection is very rare. This grass is, therefore, something more than an ordinary weed, a nuisance in cultivated fields or a temporary member of a pioneer community. It is a dominant weed that can halt and deviate the normal development of a sere in a clearing to its final mature climax. Extensive clearings invaded by *Imperata* may be held back for centuries so that they develop no further than annually burnt grasslands with scattered fire-resistant oaks. Such an arrested sere is called a subclimax—this one being a 'pyrophytic' subclimax arrested by frequent fires. Another example of a dominant weed is the prickly-pear cactus introduced into Australia for the benign purpose of making hedges. It soon dominated the countryside over enormous tracts, at one time spreading at the rate of a million acres a year. Charles Elton has discussed many other examples of such invasions.[10]

In principle, there is no difference in role and behaviour between a plant weed and a creature with legs that thrives in pioneer communities or in the cultivated fields of man. The concept of a weed is thus applicable to animals, and we should recognize *animal weeds* such as field rats and mice, garden sparrows, and apple-aphis.[11] I am stressing this concept because the generalization of the idea of a weed clarifies epidemiological thinking. Most pathogens and intermediate hosts are in the nature of weeds. I later found, as I might have guessed, that this simple idea of mine had been introduced 10 years earlier by Taylor, Vorhies, and Lister[12] who were concerned with the relation of jack-rabbits to grazing in Arizona.

Of course we use 'weed' pejoratively. We cannot help making a value-judgement and labelling a weed as 'bad'. From the point of view of the ecosystem, however, weeds are highly successful pioneers in disturbed communities. They tend to dominate, if only for a time. They colonize bare areas before any other plant can get a hold. Man should therefore feel both pride and confidence in being regarded as a dominant animal weed. This associates him with power: power to modify the ecosystem while preserving its important aspects, but also power to destroy its balance. To regard himself as a dominant animal weed without the egotistic superficial value-judgement that all weeds are bad may help man to a practical philosophy that is enormously more beneficial in the long run than, for example, the individualistic and unecological philosophy of existentialism. I am convinced that no generally adoptable philosophy is worthwhile unless it engenders genuine respect for all other forms of life, and unless the things we need to modify or adapt to are approached not only with respect but also with esteem. It is only a society that is sufficiently sympathetic to the needs and values of all living things that can afford the luxury of highly individualistic philosophies and highly self-centred behaviour among a small proportion of its members, who may be creative people, non-creative savants, or simply unsuccessful geniuses.

When I was a medical student, my professor of surgery taught me that however deeply the patient may be anaesthetized one should always operate as if he were asleep and must not be awakened. This is perhaps the best way to encourage that delicacy of technique which ensures the least bruising of the tissues left behind after surgery. I believe that a similar approach should be adopted towards all living things around us and that there is a practical advantage in being somewhat animistic. If, for reasons similar to those in favour of a kindly belief in Santa Claus, one accepts the idea that trees and animals have souls and can feel a kind of intense pain, then our attitude towards nature will be that much richer in meaning and that much more safely conservative. A man who believes he is a fully free individual with no responsibilities to the things and beings around him is a lost fellow playing his piccolo in the dark to no one in particular. His life would be much richer in content and meaning if he made himself part of Nature's orchestra, in which there are many players; his piccolo would then add something meaningful to a satisfying musical pattern.

To return to the subject of weeds in relation to scrub-typhus, it appeared to me that the infected foci of scrub-typhus so far encountered in Ceylon and the other countries were pioneer communities of vigorous and numerous weeds invading areas where the climax vegetation had been destroyed or greatly simplified. The weeds included plants (such as *Imperata* grass), the local grass- or rice-field-infesting rodents,[1] and one of the

---

[1] My colleague J. L. Harrison in Malaya found that the principal diet of many such field rodents was the numerous termites that in their turn were demolishing the stumps and logs left after clearing the forest. The implications are as obvious as they are interesting.

two dominant, widespread, and closely-related species of trombiculids, *L. deliense* or *L. akamushi*. The latter were parasitic on the rodents and were efficient vectors of the rickettsia of scrub-typhus. This rickettsial infection was boosted up in these particular conditions by the intensity of the mite-rat-mite cycles in densely populated waste land. The assumption was that if such an area were allowed to grow up into forest (or were maintained under efficient cultivation) the rodents—and therefore the mites and infection—would be reduced to a relatively and perhaps negligibly low level. Wherever *Imperata* grass took over, as it has for over a million square miles in Southeast Asia, the pioneer communities would be indefinitely arrested as subclimaxes; for the swiftly moving annual fires have only temporary effects on the small rodents and the mites. In Japan the maturing of the pioneer communities would be inhibited in a different way, by being renewed at each flooding of the river.

JUNGLE TSUTSUGAMUSHI AND 'MAN-MADE MALADIES'

The rickettsial infection in the mites must, however, have come from somewhere. Two well-known infections come to mind: yellow fever and plague. Each has a wild or sylvatic home. Each periodically escapes into situations where large populations of animal weeds boost the infection in close association with man. The yellow fever mosquito is a domiciliated animal weed, practically confined to villages and towns. Epidemic urban yellow fever is secondary to a relatively quiet endemic infection in the forest—sylvatic yellow fever as opposed to urban yellow fever. Epidemic urban plague follows outbreaks among house-rats and their fleas, all domiciliated animal weeds; but these outbreaks are originally derived from relatively quiet foci of wild plague ('sylvatic' plague although it is not in the forests) maintained in the field. It therefore seemed reasonable to assume as a working hypothesis that hyperendemic scrub-typhus was the result of intensification among field rodents to form the *yudokuchi*, but that it was derived from a relatively quiet primitive forest infection among unknown rodents and trombiculids. This latter hypothetical rickettsiosis is what I called 'jungle tsutsugamushi' infection.[4, 13] Since the rickettsia was known to be transovarially transmitted from parent mite to at least a proportion of its offspring, the infection in a given *yudokuchi* need not necessarily have come from some local sylvatic reservoir but may have been passed through a long chain of infected foci from some remote region. There was ample evidence to suggest that the two major and almost ubiquitous vectors had been dispersed very widely by birds and other animals.

On the basis of the outbreaks in Ceylon, Sumatra, and Malaya I assumed that jungle tsutsugamushi infection was unlikely to compare in importance to man with the infection among animal weeds in the field. I had been deeply impressed by the fact that no cases had been reported among those who cleared or operated in the forests. This was what I have since called the 'first prejudice': the first idea which tends to get fixed in one's mind and biases later research. It operates in the same way as the traditional idea or the 'classical' accounts that I have already mentioned (Chapter 2); but ideas conceived by oneself may tend to be held even more obstinately than ideas imprinted by tradition. They are held for different reasons.

As a malariologist, I was painfully familiar with the many ways in which man creates malaria by interfering with natural drainage and by making new breeding places for vector mosquitoes. Engineering that intensifies malaria is bad engineering. There are many appalling examples of this even when the engineers must have known the ultimate consequences of their actions. Two eminent Heath Clark Lecturers have dealt with this sad state of affairs known as 'man-made malaria'.[14, 15] At about the time I was in Ceylon, a large dam-building project not far from Trincomalee was kept so secret that even the health authorities knew nothing about it until an epidemic of infectious hepatitis was followed by an explosive epidemic of malaria, all in a remote area formerly free from malaria because it was uninhabited. (Kalra, personal communication.)

The Embilipitya focus, like the scrub-typhus foci in Sumatra and Malaya, was as man-made as anything could be; so I generalized the familiar term as 'man-made maladies'.[1] I have since been finding more and more examples and would recommend 'find the man' as a game to be played—with the warning that it can become an obsession.

THE SCRUB-TYPHUS RESEARCH LABORATORY: THE 'CIRCUS' AT IMPHAL[2]

In November 1943 at the time of the Mite Hill outbreak between Imphal and Tamu and a month before the infantry were due to dig in at Embilipitya, the Supreme Allied Commander, South East Asia Command, appealed to the British War Office for 'specialist' aid in combating scrub-typhus. The joint Typhus Committee of the War Office and Medical Research Council responded by appointing a Scrub Typhus Commission composed of Dr R. Lewthwaite, as field director, Dr Kenneth Mellanby, a well-known medical entomologist (later to be the first principal of the University College at Ibadan, Nigeria), and Charles D. Radford, acarologist. Dr Lewthwaite had been flown from the Institute for Medical Research in Kuala Lumpur to Melbourne in Australia in January 1942, together with laboratory animals infected with scrub-typhus, as an emergency measure to attempt to make a vaccine by cultivating the rickettsiae in the yolk sac of fertile eggs—this method had been very successful in the hands of Cox in the U.S.A. in the case of louse-, flea-, and tick-borne typhus, but it proved to be useless for the scrub-typhus rick-

[1] The following observation was made by me at a conference held on 3 February 1944, and published in a Ceylon Army Command report:
It is suggested that we may make a closer definition of the 'scrub' which favours typhus ... the secondary regrowth of weeds and undergrowth which invades land which has been cleared and abandoned and the natural balance of the jungle locally disturbed. It would seem that scrub-typhus may largely be a man-made disease, to be ranked with 'man-made malaria' as a consequence of human interference with natural balances.

[2] The work of this laboratory was summarized in a War Office Report[16] abstracted by Megaw.[2] Audy, Thomas, and Harrison[3] described the collecting areas and discussed the biology of the trombiculids; Audy[17] summarized some ecological aspects; Roonwal[18] described the rodents; Radford[19] and Womersley[20] described species of trombiculids; and Audy and Harrison[1] reviewed the work of the laboratories in Imphal and later in Malaya, from 1945 to 1950.

ettsiae. (The later preparation of an experimental vaccine from the lungs of infected cotton-rats was a major venture known as Operation Tyburn, but I do not propose to deal with it here—see Appendix 10.)

Members of this Typhus Commission arrived in Ceylon between April and July 1944. At that time, the malariologist for the entire Ceylon Army Command was Major S. L. Kalra, I.A.M.C., who was destined to spend a great deal of energy in scrub-typhus research. In order to set the scene, I note three events in the month of November 1944. I myself had temporarily retired from the Kabaw Valley to deal with mild hepatitis and dengue fever. An East African battalion was crossing the Bon Chaung tributary on its way to Kalewa on the Chindwin, exposing itself in this neighbourhood to several *yudokuchi* that gave rise to 85 cases of scrub-typhus, sorely taxing the unit's medical facilities just as it reached its military objective. And Kalra was on his way to Addu Atoll in the Indian Ocean. He represented the newly appointed G.H.Q. (India) Field Typhus Research Team, and was accompanied by Radford.

Scrub-typhus beset military, naval, and air-force personnel on two coral islands in the middle of the Indian Ocean. One was Diego Garcia in the Chagos Archipelago, south of the equator. This is the most far-flung lonely outpost of scrub-typhus yet known in the west. The vector mites, *L. deliense*, must have been introduced to this and other islands by birds or giant fruit-bats, sometimes hopping over 300 or more miles of ocean. A glance at the map will show that similar safaris by birds could conceivably introduce the vector and scrub-typhus through Mauritius to Madagascar and this possibility should be borne in mind. I shall refer to this in the last lecture, on the subject of new diseases.

The other infected island was Gan in Addu Atoll, Maldive Islands, some 40 miles south of the equator. Gan, an island two miles long, was an emergency air base. The first outbreak there consisted of 42 cases of scrub-typhus in Royal Marines who landed on Addu in October 1941. There were 582 in 1942, 383 in 1943, 92 in 1944, and none in 1945,[21] a declining incidence that was considerably due to the clearing and 'civilization' of the site combined with anti-rat measures. Rats had swarmed over the island and could often be seen running along telephone wires. Exposures to infection were mostly in the inner fringe of neglected coconut plantations and the native gardens. Within the former, conditions somewhat resembled those on Bat Island in the Admiralties near New Guinea in 1943.[22]

Dr Lewthwaite had had to make the difficult decision whether the three members of the Commission should restrict themselves to the propagation of known facts and countermeasures, or whether research should be started in order to fill some of the many gaps in our epidemiological knowledge. The decision was in favour of the latter.

It was decided with the military authorities that there should be a field laboratory on the Indo-Burma front, the obvious base for this being Imphal. It was not, however, until April 1945 that the Scrub-Typhus Research Laboratory, South East Asia Command, was approved as an official unit with an establishment and transport, and not until several months later that the unit actually had the transport and a complement of labour and junior assistants. Meanwhile, I had been detached from the East African Division and had spent the last two weeks of 1944 with the U.S. Typhus Commission at Myitkyina in North Burma. Here I met several renowned medical scientists and other outstanding workers, such as my friends Robert Traub, William Jellison, Glen Kohls, Gordon Davis, and the late Henry S. Fuller, who were shortly to receive international recognition for their contributions to typhus research. The first contingent of the U.S. Typhus Commission had arrived in October 1944 and, together with a General Hospital, occupied what amounted to a new village on the banks of the Irrawaddy, little more than 200 air miles east and north of Imphal. As I shall explain shortly, we attempted to work on projects complementary to those of our colleagues. In Imphal I had been joined by Harry Thomas, Royal Army Medical Corps, who had been a teacher of biology. We immediately formed a close friendship in spite of Harry's predilection for appalling puns. We lived in tents as parasites in a British General Hospital on a hillside to the northeast of

Imphal, overlooking the main airfield. The chief medical specialist there was J. N. Morris, who is now with the Medical Research Council and is the author of a popular book on the uses of epidemiology.

In the mornings, the Imphal plain would be covered in low mist and—a beautiful sight—emerging through it would be a line of shadowy trees, following a small steep-sided stream and masking the near-by tiny village of Mongjam. This village we later christened Ligula Copse because the rats there were infested by the trombiculid *Schoengastiella ligula*; but at that time, this mite was what is popularly termed 'unknown to science'. We started surveying the infected foci as well as types of vegetation, apparently uninfected areas, and various camp-sites such as that of Wingate's. For this we frequently had to thumb lifts from the transport which was always on the move along the main road. When we first got our own jeeps and trucks we were reborn, and the work started in earnest.

We were shortly joined by Kenneth L. Cockings, a civilian in the Friends' Ambulance Unit who had come to be Mellanby's assistant. Thus for the first three months of 1945 Thomas, Cockings, and I comprised a research team with the expectation at any moment of our becoming an official unit. We were shortly to develop into a very curious and heterogeneous group with much coming and going of individuals and a series of frustrations that may have been almost laughable to an outsider but seemed tragic to us.

Mellanby himself had been concerned with personal protective measures and had been responsible for an excellent film showing the manner of infection and the methods of applying dimethyl- and dibutyl-phthalate (DMP and DBP) to clothing as a protection against the mites. A close-up of a larva of *L. deliense* scampering round the head of a pin and another of larvae climbing up a sock and onto a hairy leg were to me the most memorable parts of that film. DMP and DBP, to which benzyl benzoate was shortly added, were described as mite-repellents, although they did not repel the mites, they killed them. My friend, Ronald Ribbands, had already shown me in Ceylon that DMP, then an experimental acaricide in short supply, was also a very effective leech-repellent and I profited by this.

To return to Mellanby, he was busily occupied in the important task of establishing protective measures among troops and was able to spend only a short time with the Imphal units. In early April we were joined by the Addu Atoll team: Kalra, the malariologist, pathologist, and microbiologist, and Radford, the acarologist, together with their assistants. At last, in Radford, we had someone who could identify the many trombiculids. Kalra was to stay with us throughout, but Radford left after five months, his place being taken by those he had trained, namely, Kenneth Cockings, Sgt. T. J. Lawrence, as well as W/O G. W. Ash and Sgt. W. K. Ford— Ford was entomologist to the Liverpool Museum, and it was unfortunate that he did not arrive until November. The permanent staff of I.O.R.'s arrived in mid-October. Until then we worked with the assistance of batches of Indian Pioneer Corps personnel, whose enthusiasm deserves commendation.

Three more scientists joined the team. Squadron Leader Arthur A. Bullock, D.SC., Royal Air Force Volunteer Reserve and in civilian life a botanist from Kew Gardens, who had been selected by the Medical Research Council, joined the team on 9 May and left in March 1946. H. C. Browning, PH.D., also selected by the M.R.C., arrived from the U.K. in June bringing a thousand mice. After working in Mandalay with Kalra, Browning's work was interrupted by a bout of malaria and then by a severe attack of scrub-typhus contracted in a sharply localized area in the mountains above Palel. From this camp at mile 34 from Imphal, one strain of *Rickettsia tsutsugamushi* was recovered by Kalra from Browning himself, 5 strains from two species of rat, a tree-shrew and a field mouse, 4 strains from pools of mites that included *L. deliense*, one different *R. rickettsi*-like strain from ticks.[23] It was a thorough demonstration of a *yudokuchi*! This focus was incidentally a good example of increasing endemicity of a camp-site. The camp, which included a tiny pond, was first occupied and cleared in 1943 and later abandoned. Three cases of scrub-typhus occurred among men in or around it at the end of 1944. It was then abandoned again. On reoccupation in August 1945, 22 cases of typhus occurred in the occupying unit.

Major M. L. Roonwal, PH.D., zoologist appointed from the Zoological Survey of India to the G.H.Q. (India) Field Typhus Research Team, joined both it and us on 30 June and worked particularly on the zoology of the Kanglatongbi area, 15 miles north of Imphal, in collaboration with Bullock, until he left 3 January 1946, to study his collections at the Z.S.I.[18] For two short months at the end of the year, we were fortunate to have Captain William Stanbury, R.A.M.C., whose investigations were soon interrupted. The research programme, as it was envisioned in March 1945, included a then-unappointed medical officer who would have been detached in time to take over a field team, to be based in Meiktila or some other part of Burma, and later in Malaya, to work in close contact with the troops. Unfortunately, this detachment, which would have been led by Stanbury, never materialized, and we reluctantly had to give him up to the District Laboratory in Calcutta. From May 1945 to March 1946, Captain H. C. Stewart, R.A.M.C., was our non-technical general duty officer.

CONTROVERSY ABOUT THE VECTOR

After the day's work, the evenings and sometimes much of the night were spent in discussions and arguments about the problems confronting us. One series of arguments concerned the probable vector. A member of our team had been impressed by a paper published in November 1944 by Major C. E. Cook, Australian Army Medical Corps.[24] Cook suggested that larval ticks must be considered a possible vector in Australia and New Guinea—indeed, by the manner in which he presented his argument, it almost seemed that larval ticks would satisfy all the epidemiological requirements better than any trombiculids. Therefore, argued our team member, we should proceed on the assumption that we did not even know what kind of arthropod the vector might be, whether ticks or chiggers. This argument was scientifically reasonable, provided one accepted Cook's arguments; but, even if acceptable, it was a counsel of perfection, demanding long-term studies on a very broad front. I was unable, however, to accept this viewpoint and the research programme it would entail. In a number of infected foci in Ceylon and around Imphal, we had already found *L. deliense*, known to be a vector in Malaya, Sumatra, and North India, and almost identical with the *L. akamushi* of Japan.[1] It was evidently widespread and, in many foci, a populous species. Furthermore, we were a pitifully small team, and the military problems were urgent so that we had perforce to do what I describe as poking open buds in order to get flowers. Therefore, my argument was that we should assume that *L. deliense* was the primary local vector and study it as intensively as possible. During the course of these studies, we could plan field-work to detect other potential vectors. In the end, at least we would have added to our knowledge of an acknowledged vector, whatever else we did. Basing our work very broadly on Cook's hypothesis, however, we might readily work hard for a year with only negative results. Furthermore, Cook's argument was based on some unwarranted assumptions.

This argument, heated at times, did not affect either our research or my decision that we should go our own ways in the field because I felt convinced that *L. deliense* would force itself upon us to the complete exclusion of any larval ticks.

Since we are concerned in these lectures with the development of ideas and the process of planning research, or rather, with hindsight with respect to these matters, it is necessary to inquire into Cook's reasons for reaching his conclusions. He had encountered scrub-itch among Australian troops in the Atherton tableland of Queensland and in New Guinea, and he had undoubtedly been influenced by the work of Carl Gunther in New Guinea.

[1] *L. deliense* was first seen, but not identified, not far from Imphal as long ago as 1914, when J. M. D. Mackenzie of the Indian Forest Service collected eight specimens of the Burmese tree-shrew *Tupaia belangeri siccata*, from the upper Chindwin area (Kinday) and the Chin Hills, in areas now known to be hyperendemic with scrub-typhus. He noted that 'All shot were found to have a curious orange patch on the rump, apparently eggs producing 6-legged, round animals about half as big as a pin's head.'[25] These were chiggers, almost certainly *L. deliense*, described in Sumatra eight years later. This was four years after the description of *L. akamushi* by Brumpt. Mackenzie's mistake was a reasonable one. A colony of engorged attached chiggers looks very much like a mass of insect eggs such as one may find on a leaf. Had Mackenzie collected his little round animals for an acarologist to study, our knowledge of the possible geographic distribution of scrub-typhus might have been accelerated. Tree-shrews are important hosts of *L. deliense* in many places.

Gunther, in 1940,[26] reported that about 90 per cent of some 2000 chiggers collected by him in New Guinea were *Eutrombicula wichmanni* (misidentified as *Trombicula minor*). This species, a common cause of scrub-itch, is essentially a bird-chigger, and he collected ten times as many birds as mammals.[1] We should contrast Gunther's findings with those of Heaslip (1941) who reported that 90 per cent of some 2000 chiggers collected from mammals were *L. deliense*.[28] Gunther's approach to the epidemiological facts, especially to the fact that scrub-itch and scrub-typhus do not often coincide, was decided by his biased collection, together with a facile assumption that any attack by any chigger would be accompanied by local reaction, i.e., scrub-itch. He decided that the primary vector in Queensland and New Guinea was *E. wichmanni* ('*T. minor*'), but that only a very small proportion of these chiggers were infected, this proportion being those few that had fed on rodents or rodent-like marsupials, which animals had in turn been infected by a species of chigger such as *T. vanderghinstei*. (This last-named species was, in fact, *L. deliense*, which he had found only in relatively small numbers through biased collecting.)

Cook unfortunately started with the false premise that 'the only certain evidence that a species of mite will attack man is its production of itch'. Therefore, he was forced to conclude that 'there are a number of contradictions to any species [of chiggers] being the vector of scrub typhus'. For example, not only did scrub-itch and scrub-typhus rarely coincide, but scrub-itch, in New Guinea, was peripheral on the body, especially on feet and legs, while scrub-typhus eschars were on the trunk; scrub-itch lesions were multiple, eschars usually solitary; 'mite bites are irritable' but the eschar is painless and usually overlooked by the patient. (Gunther had suggested that the locally multiplying rickettsiae destroyed nerve-endings and anaesthetized the site.) It therefore appeared to him that 'the vector in endemic foci is sufficiently infrequent to insure only one infective bite per man, but numerous enough to infect an appreciable number of men in the same locality at any one time'.

[1] Consett Davis, by collecting only reptiles, recorded four species of *Eutrombicula*.[27]

DOUBTS ABOUT RELATIONS TO VEGETATION

To be presented with the whole field of scrub-typhus epidemiology and the urgency of wartime casualties—to be told to get moving and do something quickly in such circumstances—is a sobering responsibility. It is easy to become buried in academic pursuits and equally easy to achieve nothing by hasty pursuit of an applied objective. Furthermore, the U.S. Typhus Commission at Myitkyina in North Burma was much better staffed with outstanding scientists and much better equipped than ourselves, and it had already been energetically carrying out surveys. Therefore, although duplication of work is by no means necessarily wasteful, in these circumstances anything we might simultaneously do would probably be duplicated much better by them and our contribution thus would hardly help the combined efforts of the Allies to prevent scrub-typhus in troops. We should do complementary work, if possible.

I was already convinced that the distribution of scrub-typhus must somehow be related to the vegetation. This conviction was based on (a) the knowledge that populations of small mammals are related to vegetation, and therefore so would be their parasites such as chiggers; (b) the localization of chiggers and infections to 'mite-islands' and 'typhus-islands' (*yudokuchi*), which would hardly be fortuitous; (c) the suspicion that hyperendemic areas were largely man-made, each mite-island being a part of a complex of weeds; and (d) the (mistaken) belief that the post-larval stages of the chiggers might feed on plant juices from the superficial mat of roots. It therefore seemed wise to pursue these ideas, that is, to see if indeed any relationship could be found between scrub-typhus and either particular types of vegetation or particular features of the vegetation pattern. Almost all the countryside ahead of our troops was well covered by air photographs, and we were using these for malaria surveys. The troops themselves were performing a large-scale experiment by swarming over the countryside. I decided that we should concentrate on air photography, supported by the necessary work on the ground. By studying air photographs we could perhaps detect potentially dangerous areas ahead of troops and warn them accordingly (at that time I was naïve enough to think that a military commander could afford to listen to such tentative warnings). The main purpose, however, was to use air photography as a form of armchair exploration, as an epidemiological research tool.

All the outbreaks I had encountered or learnt about so far seemed to be fitting into a pattern. This was, of course, greatly influenced by my particular experience, especially those earliest encounters that tend to establish one's first prejudices. Some colleagues in the U.S. Typhus Commission at Myitkyina, with more scientific background than I, had had different experiences and therefore were much more cautious in interpretation. Late into many nights we discussed differing views. In order to follow my own beliefs freely I made a decision that I have often thought about since. I knew that the dogged pursuit of an idea against all setbacks requires faith; and faith is blind. I was convinced that if there is order in nature there must be some relationship between vegetation pattern and the distribution of dense populations of small mammals and vector mites. Nevertheless, the situation was very complex. I doubted that jungle tsutsugamushi in the forest would be important to man, but this idea was then being shaken by what appeared to be an outbreak in 'virgin' forest between Mawlu and Pinwe.[29] Therefore, I resolved to be unscientifically obstinate and to refuse to be put off by 'confusing accounts' unless they were supported by the actual identification of infected chiggers on the ground. In other words, the onus of disproof, up to the hilt, must be on others. There was, of course, always the possibility of local alternative vectors, but *L. deliense*, and possibly *L. akamushi*, seemed to be almost ubiquitous and the dominant species over a vast geographical area. Unscientific as it was, I think that this deliberate obstinacy served its purpose. It preserved enough faith to keep my nose to the grindstone in a single-minded pursuit. I have wondered if the same process operated with Ludwig Anigstein in Malaya when he equally obstinately continued pursuing forms of the proteus and other bacilli.

The *modus operandi* was simple. The troops themselves had made a thorough survey by combing the countryside for infected areas. Many outbreaks could be pinpointed, that is, traced back to definite foci, and these, together with uninfected camp-sites, were studied by air photographs. For those who are unfamiliar with the precision of air photography I need only mention that by the use of the stereoscope on photographs taken from an altitude of 40 000 ft, it is easily possible to see that the roadside drain is at a distinctly lower level than the surface of the road and to count the branches on a dead shrub. By this means, even patterns of human behaviour in relation to land-usage and to long-abandoned villages or cultivation complexes can be traced backwards in time over vast areas. Armchair exploration with air photographs can be even more exciting and exhilarating than exploration on the ground, with which it must however be combined.

Meanwhile, the usual investigations by trapping animals and collecting mites off the ground were carried out in many representative areas.[3, 23] Many months before, while we were advancing down the Kabaw Valley, my working hypothesis had been supported by finding *L. deliense* on the ground in a grassy swathe cleared through forest under a telephone line, but not in the adjoining forest. During the course of pinpointing a large number of outbreaks and studying the terrain around the infected sites by air photography or ground inspection, the relationship to vegetation became increasingly confused. The picture suddenly became clearer when I realized (a) that there seemed to be more risk at the edges of grassland than in the centres—an observation supported by some reports from New Guinea and Australia—and (b) that there was also special risk associated with what I called the 'hedgerow type of feature'. The latter, which comprised hedgerows and fingers of gallery forest following streams and ravines out into open grassland, also seemed to comprise two forest-scrub edges which met in the middle. I called these and the forest edges 'fringe habitats' because their life-form and denizens were different from those of either the forest or the scrub. On returning to civilization in August 1946, I learned with great excitement that game-managers in particular had long recognized such 'edge-effects'.[30] Special features that affect animal life are to be found where one vegetation type joins another.

It soon became apparent that most of the typhus-islands or

vudokuchi were located in three types of terrain within the patchwork of mixed vegetation which follows shifting cultivation, deforestation, and settlement. Foresters in the Philippines have a picturesque name for this patchwork, *parang*-vegetation, because it follows the use of the *parang* or machete. In brief, the three types of terrain are:

A. *Man-made waste land*

(1) *Rural waste land*—overgrown clearings, the result of temporary shifting cultivation and known as *ponzos* in Burma, especially if these are restricted in area and also especially by the edges or near moist places such as hollows or stream banks.

(2) *Domestic waste land*—neglected weedy patches of land such as are found around habitations and even in big towns. A classical example is the patches of grass within gardens in the Pescadores Islands. In about 1952 scrub-typhus became a serious problem in Jamshedpur in India.[31] This is a very dry region and infected areas were almost entirely confined to the watered gardens of the more prosperous people—a domestic situation reminiscent of the intense infestation in gardens by pest-chiggers in Europe and the United States, where the chiggers may be largely supported by birds attracted to the gardens.[32]

(3) *Neglected native gardens or plantations, and abandoned villages*—areas with a relatively plentiful food supply. In some of these, the risk of infection may be high indeed. Abandoned villages are not uncommon in remote areas, and among the various causes for their abandonment one may mention the occasional flowering and fruiting of bamboos—a dramatic event followed by famine.[1]

B. *Water-meadows*

Water-meadows are grassy, but not generally swampy, river and stream banks or islands. Small water-meadows may be found along meandering streams in virgin forest. Others are formed by periodical flooding. Over 100 men camped in the

---

[1] See Appendix 9 for a note on the mass flowering of gregarious bamboos.

Kabaw Valley some 50 miles north of Tamu. Seven men contracted scrub-typhus: they were among officers and their batmen who camped on a small water-meadow by the meandering Wetyu stream—an appropriate name. The men who camped and patrolled in the forest escaped.[33] The classical Japanese *yudokuchi* are similar to water-meadows, whence we get one of the names for the vector mites, *shima-mushi,* or island-mite.

c. *Fringe-habitats*[17, 34] *and Secondary Scrub*

These include a range of features from a simple bushy hedgerow to the strips of gallery forest which are commonly left following water courses or ravines, as well as the scrub-forest fringe. Not only do these appear to be remarkable sanctuaries of small animal life, but they tend to conserve and accompany moisture, thus allowing the chiggers to continue to flourish into the dry season. We were able to get experimental confirmation of the increase in chigger population in a narrow band along such fringe-habitats as well indeed as along the edge of our own camp. In India and Burma *L. deliense* occurs profusely in open grassland as well as in fringe-habitats which may be well-wooded. In the equatorial region the distribution appears to be somewhat different: where it occurs, *L. akamushi* seems in these regions to be absolutely restricted to grass and the hosts found there (such as the rice-field rat and quail in Malaya[35]) while *L. deliense* occurs much more widely and especially in fringe-habitats and secondary scrub or forest with leaf-litter,[36–8] even occurring in deep forest but in relatively very small numbers.[1]

Safe rules cannot be laid down for avoidance of infection in such conditions but the experienced worker can nevertheless hazard good guesses as to potentially dangerous places, either by direct inspection of the terrain or by studying air photographs. One rule-of-thumb that emerged from these studies at Imphal was that the risk should increase with the number of blocks of vegetation of different types in unit area. This was a simple way of measuring the extent of edge or of fringe-habitats. In some places, a single square mile might contain over 50 blocks of vegetation easily detected in air-photographs, or over 20 miles of edge. The blocks of vegetation usually represented numerous old clearings in various stages of regeneration (Figs. 11, 12).

It is noteworthy that while these investigations were going on we received a translation of a Japanese report on this infection which was giving so much concern to the armed forces on both sides. It is an astonishing fact that the disease had not been recognized by those Japanese investigators: they gave it the tentative name of Burmese eruptive fever, or in one area *tenta* fever (*tenta*, Burmese for a bushy swamp), and speculated that it might be transmitted by fleas. What held these competent clinicians and epidemiologists from diagnosing a disease familiar at least by repute to every Japanese physician? I am sure it was the dead hand of tradition, the clinical and epidemiological mental picture of 'classical' tsutsugamushi disease along the rivers in northwest Honshu. Even the lessons learnt by the Japanese themselves in Formosa were apparently forgotten. In the last Lecture I shall suggest that this vivid picture of the classical disease also held back Japanese recognition of forms of tsutsugamushi disease transmitted by other species of chiggers over the whole length and breadth of the Japanese islands.

REFERENCES

1. AUDY, J. R. and HARRISON, J. L. (1951). A review of investigations on mite typhus in Burma and Malaya, 1945–1950. *Trans. roy. Soc. trop. Med. Hyg.* **44**, 371–404.
2. MEGAW, J. W. D. (1948). Scrub typhus investigations in South East Asia (Abstract). *Trop. Dis. Bull.* **45**, 62–70.
3. AUDY, J. R., THOMAS, H. M. and HARRISON, J. L. (1953). A collection of trombiculid mites from Manipur and Lower Burma, 1945–1946. *J. Zool. Soc. India* **5**, 20–40.
4. AUDY, J. R. (1949). A summary topographical account of scrub typhus 1908–1946. *Bull. Inst. med. Res., Malaya* **1** (new series), 1–86.
5. MIYAJIMA, M. and OKUMURA, T. (1917). On the life cycle of the 'Akamushi', carrier of Nippon River Fever. *Kitasato Arch. exp. Med.* **1**, 1–14.
6. MEHTA, D. R. (1937). Studies on typhus in the Simla Hills. VIII. Ectoparasites of rats and shrews with special reference to their

PLATE I

FIG. 11. Oblique air-photograph, mountainside on Indo-Burma border, showing the advance of shifting cultivation and human settlement. Note the settled villages and terraced rice-fields under irrigation at the bottom of the mountain; the large area of shifting-cultivation showing grasslands (pale) with relict strips of gallery forest (dark) following the streams and ravines; and the unspoilt forest near the top. Royal Air Force copyright.

Plate II

Fig. 12. Vertical air-photograph, remote area of north-west Burma (roughly 9509°E. 2606°N.) about 25 miles S.W. of Shingbwiyang. Illustrating clusters or 'constellations' of shifting cultivation. Recent clearings have been accentuated in white and the remote villages also picked out and marked V. The area is divided topographically into six areas (N O P, S T, and U, being separated by rivers). N: rugged afforested mountains of the Sangpang Bum range. P: crumpled and steep-sided hills covered with patches of grass, scrub, forest, with gallery forest belts along watercourses, and extensive secondary bamboo tracts. T: a similar section with steep ridges—very difficult country to travel through. Note the fresh clearings c.2 separated by nearly a mile of rugged country from another constellation c.1 shown in black, made several years before and now covered with scrub (village V6 responsible). O, S, and U are less rugged and contain much less residual forest, *e.g.*, under 15 per cent of S is forest, all in the form of narrow gallery forest or small patches. X is the end of a large constellation of fresh clearings belonging to a village, off the photograph, as large as V1 to V6 together.

This illustrates features presumably concerned with the large-scale patchy distribution of scrub-typhus. The six areas each have distinct topographical individuality, reflected in settlement and land-usage. Each is separated from others by geographical features resembling those which on a larger scale separate Burma from Assam or the Shan States. Many of these barriers are also ecological, to a limited extent isolating breeding groups (demes) of animals. The constellations of clearings represent space-time events that greatly influence the vegetation and wild animal populations over restricted and isolated areas, presumably leaving behind residual changes in parasite patterns. Royal Air Force copyright.

possible role in the transmission of typhus. *Ind. J. med. Res.* **25**, 353–65.

7 OUDEMANS, A. C. (1912). Larven von Thrombidiidae und Erythraeidae. *Zool. Jahrbuch. Suppl.* **14**, 45–62.

8 WHARTON, G. W. (1946). Observations on *Ascoschöngastia indica* (Hirst, 1915) (Acarinida Trombiculidae). *Ecol. Monogr.* **16**, 151–84.

9 JAYEWICKREME, S. H. and NILES, W. H. (1946). Successful feeding experiments with an adult trombiculid mite (order Acarina). *Nature, Lond.* **157**, 878.

10 ELTON, C. S. (1958). *The Ecology of Invasions by Animals and Plants.* London, Methuen & Co.

11 AUDY, J. R. (1963). Man and the land *in* F. LEYDET, Editor, *Tomorrow's Wilderness*, pp. 101–60 San Francisco, Sierra Club.

12 TAYLOR, W. P., VORHIES, C. T. and LISTER, P. B. (1935). The relation of jack rabbits to grazing in southern Arizona. *J. Mammal.* **37**, 358–70.

13 AUDY, J. R. (1956). The role of mite vectors in the natural history of scrub typhus. *Proc. 10th Congr. Ent.* **3**, 639–49. [See also AUDY[4]]

14 HACKETT, L. W. (1937). *Malaria in Europe. An Ecological Study* (Heath Clark Lectures, 1934). Oxford University Press.

15 RUSSELL, P. (1955). *Man's Mastery of Malaria* (Heath Clark Lectures, 1953). Oxford University Press.

16 AUDY, J. R. and others (1947). *Scrub Typhus Investigations in South East Asia. A Report on Investigations by the G.H.Q. (India) Field Typhus Research Team, and the Medical Research Council Field Typhus Team, based on the Scrub Typhus Research Laboratory, South East Asia Command, Imphal.* (Mimeographed, Illustrated.) Great Britain, War Office. [See Appendix 5 and MEGAW[2]]

17 AUDY, J. R. (1947). Scrub typhus as a study in ecology. *Nature, Lond.* **159**, 295–6.

18 ROONWAL, M. L. (1949). Systematics, ecology and bionomics of mammals studied in connection with tsutsugamushi disease (scrub typhus) in the Assam-Burma war theatre during 1945. *Trans. natl. Inst. Sci. India* **3**, 67–122.

19 RADFORD, C. D. (1946). New species of larval mites (Acarina Trombiculidae) from Manipur State, India. *Proc. zool. Soc. Lond.* **116**, 247–65. Also, (1946) Larval and nymphal mites from Ceylon and the Maldive Islands. *Parasitology* **37**, 46–54.

20 WOMERSLEY, H. (1952). The scrub-typhus and scrub-itch mites (Trombiculidae, Acarina) of the Asiatic-Pacific region. *Rec. S. Aust. Mus.* **10**, 1–435. Illustrations separately bound.

21 KALRA, S. L. (1947). Addu atoll (Maldive islands), its people and its important diseases. *J. Ind. Army med. corps* **3**, 137–41.

22 PHILIP, C. B. and KOHLS, G. M. (1945). Studies on tsutsugamushi disease (scrub typhus, mite-borne typhus) in New Guinea and adjacent islands. Tsutsugamushi disease with high endemicity on a small South Sea island. *Am. J. Hyg.* **42**, 195–203.

23 BROWNING, H. C. and KALRA, S. L. (1948). Scrub typhus subsequent to

'Fulton' vaccine and investigation of infected site. *Ind. J. med. Res.* **36**, 279–90.
24. COOK, C. E. (1944). Observations on the epidemiology of scrub typhus. *Med. J. Aust.* **2**, 539–43.
25. WROUGHTON, R. C. (1916). Mammal survey of India, Burma and Ceylon: Report No. 25: Chin Hills (coll. by J. M. D. Mackenzie, I.F.S., 1914–15). *J. Bombay nat. Hist. Soc.* **24**, 765.
26. GUNTHER, C. E. M. (1940). A survey of endemic typhus in New Guinea. *Med. J. Aust.* **2**, 564–73.
27. Consett DAVIES coll.: see WOMERSLEY[20].
28. HEASLIP, W. G. (1941). Tsutsugamushi fever in North Queensland, Australia. *Med. J. Aust.* **1**, 380–92.
29. See AUDY,[4] page 26.
30. LEOPOLD, A. (1933). *Game Management* (Chapter v). New York, Scribner.
31. SWAMY, T. V. and DUTTA, B. B. (1953). Epidemiology of XK typhus in Jamshedpur. *Ind. med. Gaz.* **88**, 522–5.
32. TUXEN, S. L. (1950). The harvest mite, *Leptus autumnalis*, in Denmark. *Entomol. Medd.* **25**, 366–83.
33. SAYERS, M. H. P. and HILL, I. G. W. (1948). The occurrence and identification of the typhus group of fevers in southeast Asia. *J. roy. Army Med. Corps* **90**, 6–22.
34. AUDY, J. R. (1965). Types of human influence on natural foci of disease *in* B. ROSICKÝ and K. HEYBERGER, Editors, *Theoretical Questions of Natural Foci of Disease*, pp. 245–51. (Proceedings of a Symposium, Prague, 26–29 November 1963.) Prague, Publ. House of Czechoslovak Academy of Sciences.
35. AUDY, J. R. (1956). Trombiculid mites infesting birds, reptiles, and arthropods in Malaya, with a taxonomic revision, and descriptions of a new genus, two new subgenera, and six new species. *Bull. Raffles Mus., Singapore* **28**, 27–80.
36. HARRISON, J. L. (1956). The effect of grassfires on populations of trombiculid mites. *Bull. Raffles Mus., Singapore* **28**, 112–19.
37. GENTRY, J. W., CHENG, S. Y., and PHANG, O. W. (1963). Preliminary observations on *Leptotrombidium* (*Leptotrombidium*) *akamushi* and *Leptotrombidium* (*Leptotrombidium*) *deliense* in their natural habitat in Malaya (Acarina: Trombiculidae). *Am. J. Hyg.* **78**, 181–90.
38. HUBERT, A. A. and BAKER, H. J. (1963). Studies on the habitats and population of *Leptotrombidium* (*Leptotrombidium*) *akamushi* and *L.* (*L.*) *deliense* in Malaya (Acarina: Trombiculidae). *Am. J. Hyg.* **78**, 131–42.

5

OLD AND NEW HORIZONS

THIS is the last Lecture but there is still so much to tell. I have decided not to develop a single theme, but, at the risk of appearing to jump from one subject to another erratically, to take up several.[1] Each illustrates a step in our understanding of processes in nature relevant to preventive medicine. Some steps take us into areas which are still being explored, where the new horizons are still far away.

CHANGING PATTERNS OF DISEASE

The patterns of disease in human communities are constantly changing, reflecting changes in the exploitation of the countryside, the magnitude and nature of the usual migration from rural to urban areas, the abuse of chemicals (such as pesticides, alcohol, tobacco), the development and nature of pollution (of air, water, food), the socio-economic development of the community as well as its age-structure and the psychological state of its component humans, who respond to social stresses in complex ways—and many other changes. Lessons relevant to such changing patterns can be learnt from the course of evolution and establishment of scrub-typhus, even though some elements in the evolution of this infection must be conjectural.

In the third Lecture I presented a working hypothesis that all the fevers of the typhus group might have originated in rickettsiae among trombiculid mites and their hosts rather than among ticks. This hypothesis can be tested only after much labour and time. Alternative hypotheses concern the ultimate origin of these rickettsiae and whether they are monophyletic or polyphyletic. There is little doubt, however, about an acarine rather than an insect origin, or about the

---

[1] See Appendix 10 for notes on prevention and treatment of scrub-typhus, a section which has been removed from this chapter.

adaptation of the rat-flea-borne *Rickettsia typhi* to human body-lice, with the emergence not only of epidemic typhus but of a group of rickettsiae pertaining to man himself and his lice: *R. prowazekii*, *R. quintana*, *R. rochalimae* (or some other non-pathogenic form), and perhaps others in the human louse. This adaptation had a tremendous effect on man. A relatively unimportant sporadic infection derived from rats became a terrible epidemic disease. A similar adaptation has taken place in spirochaetes (*Borrelia*) from ticks to human lice; epidemic relapsing fever emerged in this way from endemic tick-borne relapsing fever.

Evidence suggests that there is a widespread tsutsugamushi-disease-like rickettsiosis in chiggers in the region roughly bounded by Japan, West Pakistan, archipelagoes in the Indian Ocean, and tropical Queensland. What we know of this rickettsiosis is almost entirely confined to those rickettsial strains that are recoverable by the usual mouse-passage. Everyone who has worked in the field has been certain that he has detected other strains but has lost them before they could be properly identified. Our techniques of recovery and isolation must be improved if we are to learn more about the underlying rickettsiosis.

In innumerable places, the wild infection among the 'jungle-tsutsugamushi' has been intensified by local populations of chiggers that will bite and infect man if given the opportunity. These chiggers seem always to belong to the essentially ground-dwelling genus *Leptotrombidium* (subgenus *Leptotrombidium*). In some of these places, man has created the conditions that have encouraged the boosting of infection.

Two closely related chiggers, *L. deliense* and *L. akamushi*, have emerged as vigorous species eminently adapted to man-made conditions in which field-rats densely populate abandoned clearings and the adjacent forest fringes. These two vectors have been widely dispersed, especially by birds, until they are now the primary vectors over the whole range of distribution of scrub-typhus. In most places they are the sole vectors. Of the two, *L. deliense* has by far the wider and more thoroughly established distribution. *L. akamushi* seems to have evolved from a common stem with *L. deliense* in some region towards the east, possibly in one of the islands in the Western Pacific and quite possibly in northwest Honshu itself. Both these species are still being dispersed, *deliense* having a lead over *akamushi*. Their possible subspeciation has not been studied.

It is therefore possible to discern the evolution of a potential pathogen (*Rickettsia tsutsugamushi*), the emergence of *yudokuchi* (areas potentially dangerous to man), the introduction of one of the vigorous vectors and infection to hitherto unaffected places, and local changes in response to man's utilization of the land. There will have been times when scrub-typhus has appeared as a 'new' disease.

CHANGING PATTERNS: 'NEW' DISEASES

An apparently new disease may be only newly recognized, but it may be newly emergent, newly introduced, or newly evolved. There are also diseases whose aetiology has only recently been decided: these newly elucidated diseases do not concern us here.

Examples of *newly recognized* diseases, already existing but overlooked, are the various local forms of scrub-typhus discovered in Japan since the war. Following an outbreak in military forces in a wholly unexpected locality on the slopes of Mount Fuji[1] in 1946, a very thorough survey was made over the whole of Japan.[2] As a result, several distinct epidemiological types of scrub-typhus were discovered,[3, 4] one of the most interesting being the so-called Shichito Fever or winter scrub-typhus, transmitted by *L. scutellaris* especially in the Shichito (seven islands) in the Bay of Tokyo.[5]

A distinct form of 'recognition' of a disease is of course the first encounter with an enzootic focus, that is, a nidus of an infection which has been silently circulating among animals and associated vectors. It was the encountering of such foci of tick-borne encephalitis during the opening up of the Siberian hinterland that started the late E. N. Pavlovsky and his colleagues on the development of their 'doctrine of nidality'. [6, 7]

A *newly emergent* disease is one that has existed on a small scale, often undetected, but has become prevalent as a result of changing circumstances. The outbreaks of scrub-typhus that appeared in Jamshedpur, India[8] following the development of favourable conditions in watered gardens are an example. Others are man-made malaria, epidemic urban plague and yellow fever (when they first appeared), and Western equine encephalitis following irrigation of the San Joaquin Valley in California. Paralytic poliomyelitis, however, is an example of another sort.

Newly introduced diseases are difficult to exemplify with scrub-typhus except perhaps for the presence of one or other of the major vectors as introduced species in oceanic islands and tropical Queensland. *L. deliense* is certainly an introduced species in Australia, where it and scrub-typhus are confined to a very small region that receives a tropical ration of rainfall and is roughly bounded by the 60 in. isohyet. There are however a number of places farther south where conditions could readily support *L. deliense* and scrub-typhus, once introduced. *L. deliense* and the rickettsia have been introduced to a number of oceanic islands stretching between India and Madagascar, the southernmost limit so far known being the Chagos Archipelago (Diego Garcia). It is evident that birds (and possibly flying-foxes) may carry *L. deliense* over several hundred miles of ocean. Perhaps it is only a matter of time before this vector becomes established in Madagascar, for what little is known of flight-lanes of birds in the southern Indian Ocean suggests that birds may link these islands to each other and to the mainland.

The potentialities of the slow spread of scrub-typhus are by no means exhausted. Not only may it spread to new areas by introduction, but wherever it exists there are opportunities for further entrenchment.

Newly evolved diseases are genuinely 'new'. A recent example of these appears to be the mosquito-borne dengue-group haemorrhagic fevers of Southeast Asia.[9] At one time, endemic flea-typhus was a newly introduced disease to man, brought to him by the rats he encouraged in his homes. Later, and within relatively recent times, epidemic louse-typhus and, in due course, trench fever were newly evolved. Rickettsialpox, as it was encountered in Boston, was probably a newly emergent disease but it probably evolved in recent historical times in relation to house-mice introduced ultimately from Asia.

REVISION OF IDEAS ABOUT RATS

Before introducing Pulau Jarak, one of many islands pullulating with rats, it would be appropriate to mention some of the confused thinking that has surrounded rodent control in relation to scrub-typhus.

Since scrub-typhus is based on rats and their parasites, what assumption could be more natural than that rats should be destroyed as a protective measure? This seemed an obvious recommendation and is still to be found in the literature, but there is a great difference between anti-rat hygiene as a long-term preventive measure and an anti-rat campaign as an emergency measure, for the latter will usually *increase* the risk of human infection.

The first question we had to answer in Imphal was this: Even though chiggers normally take only one feed, will they reattach to another animal if a feed is interrupted when the host is killed? This can easily be tested: Glen Kohls in the U.S. Typhus Commission in Burma simply stitched the severed ears from a heavily infested rat to the ears of a clean rat, and we at Imphal suspended a freshly killed, heavily infested rat by the tail above the head of a clean rat imprisoned in a wire-mesh cylinder. We both found that a few larvae of *L. deliense* would reattach, presumably because they had had less than a threshold feed. Hygiene personnel were advised to burn every trapped rat and to treat the ground where it had lain with insecticide or in some other suitable way.

The essential question, however, concerned the proper way to destroy the rats and the conditions under which this was even desirable. Rats are the favoured maintaining hosts of the chiggers and no animal is more expert than they in mopping up hungry larvae. Furthermore, as J. L. Harrison showed,[10] each rat tends to build up a personal infestation in its runs. This was detectable in 2 ways during the course of experiments when rats were trapped, marked, released, and repeatedly retrapped. Each individual rat would maintain the same sort of pattern of chigger-species time after time, while a neighbouring individual would regularly have a slightly different pattern (I should explain here that rodents usually support several species of chiggers, the record being held by *Rattus bowersi* in Malaya with over 40 species of chiggers and no less than 20 species on one single individual). Also, from time to time a marked rat would be kept in the laboratory for a few days before being released in its home-range, but would be trapped again a day or so after. In these cases, the rat would have that extra number of chiggers on it which might be expected if it maintained a 'personal' infestation in its runs and burrow, and if the newly hatched chiggers had been waiting for their master to return.

Kill the maintaining hosts and every day more hungry larvae hatch out and await a chance encounter with some alternative host. This may be one of the factors that account for some instances in which scrub-typhus has started some time after the establishment of a camp:[11] rodent control might have been established in the periphery. (See Appendix 8.)

Nevertheless, advice to destroy the rats still persists. This is helpful if it is part of a long-term hygienic programme and is combined with treatment of the soil. Otherwise, for short periods of exposure it is better to encourage the rats! Although I still have to describe conditions on Jarak Island, I may mention that 6 of 8 collectors visiting it in 1932 acquired scrub-typhus. On recent visits starting in 1952 there was not a single casualty. The 1932 party amused themselves in the evenings by shooting rats as they came into the camp-site, collecting carcasses near each bedside so that competitive scores could be counted as the rats were thrown into the sea the following morning. Perhaps a thousand or more rats were shot in this way in a few days. This would have greatly increased the risk of infection to the party. On the later expeditions, rats were encouraged in the camp in order to watch their behaviour, especially interactions between rats and between rats and the huge land-crabs. These two ways of dealing with the rats were ideally suited to increase, or on the latter occasion, decrease, the risk of human infection.

Cases of scrub-typhus among Royal Air Force personnel

and their families were traced in 1956 to waste land in the middle of Changi camp, Singapore. This area was circumscribed by a circular road and the high incidence of *L. deliense* and scrub-typhus was confined by this ecological barrier. The plan of control required that rats heavily infested by infected chiggers should not be driven out by the control measures into the surrounding scrub. Therefore, some preliminary mark-release experiments were necessary to find out what the rats were doing. The medical officer in charge of investigation and control was soon ordered by higher authority to stop this nonsense of actually releasing the rodent arch criminals. As a result, an opportunity to learn something important was lost, and infection might have been driven out into the surrounding scrub (but, as it happened in this case, probably was not). I mention this as an example of the tribulations of those called in for specialist aid; it will strike a chord with every epidemiologist who has conducted field campaigns. On the other hand, the 'higher authority' is usually painfully familiar with the unworldliness of the scientific investigator, who may have no conception of practical matters and who may be more entranced with the investigation than with the control. I have been guilty of this myself and well remember forgetting for a while the need for malaria control whilst working overtime on an entrancing mosquito study in the Chittagong hill-tracts in 1944.

The clarification of our ideas about the relationship of rats to numbers of chiggers and their patterns of distribution in space is one minute part of a rapidly growing understanding of 'place diseases': the distribution of zoonoses and the influence of them on man's activities.[12, 13]

JARAK: ISLE OF RATS[14, 15]

Those who come down the Malacca Straits between the Malayan Peninsula and Sumatra pass an uninhabited green island rising some 500 feet out of the sea. This is Pulau Jarak (*pulau* = island), shown as Ch'en Kung Hsü—Master Ch'en's Island—in Chinese maps of the fifteenth century. Master Ch'en Hsü was a powerful pirate chief who had his headquarters at Palembang at Sumatra at the end of the fourteenth century. His island in the middle of the Straits commands a view across the channel to the mainland on both sides, and its name suggests that it was favoured by Ch'en as a base for intercepting shipping. Although one of the Malay meanings of *jarak* is intervening space, the true origin of this name is apparently unknown. The Chinese fishermen often call the island Pak Ku, the northern turtle, and this is what it looks like, a great green turtle half a mile long rising straight out of the sea from a fringe of enormous rocks and covered in dense forest.

In 1932 a collecting expedition from the Raffles Museum (now National Museum), Singapore, visited Jarak. The party consisted of about 8 people, 6 of whom contracted scrub-typhus as a result of the visit. For this reason, a second party of 10 members, including John Wyatt-Smith, forest botanist, and J. L. Harrison and myself from the Scrub-Typhus Research Unit at the Institute of Medical Research at Kuala Lumpur, visited the island for 10 days in January 1950 and some of us returned at rare intervals thereafter whenever we could afford the funds and the time. Our guide was a Dyak museum collector, Ben Ensoll, an unforgettable character who was one of the victims of scrub-typhus on the 1932 expedition.

During the daytime the island was silent save for the occasional call of birds and the infrequent, startlingly loud call of the giant tuck-too gecko, punctuating the wash of the waves against the huge boulders down below. In the evening the island woke up. Large land crabs measuring 5 or more inches across the carapace sidled out of holes from under tree roots and among the rocks. The rats came out in force until the silence of the night was broken by the noise of their frequent encounters, by the chattering of one rat teaching another a lesson. These rats behaved as if they were completely tame. It was possible to tickle rats with the forefinger, and for the whole group of us to eat our evening meal with 10 to 20 rats feeding unperturbed in the same tent. Harrison later estimated the population of rats to be at least one per 10 square yards or perhaps 50000 on the island. The overcrowding was reflected by the state of health of the rats: one in 3 had bald patches, practically every rat was wormy, and most of them had damaged tails and ears as a result of their incessant fighting. Old males would commandeer a piece of coconut and drive others off in a very amusing way by kicking out backwards and sideways with the hind foot, without even looking up. All the rats were freely infested with the scrub-typhus vector, *L. deliense,* and we were able to confirm the presence of infection among them.

After only one day and night of exploration of the island, we agreed unanimously that the flora and fauna were in no way representative of the mainland on either side. In other words, and to our great surprise, this was not a continental but an oceanic type of island which must at one time have been completely bare. It was too high for us to account for this by changes in sea level, but M. W. F. Tweedie later produced what must be the explanation.[16] Lake Toba in Sumatra was a great mountain until it blew off its top in a gigantic volcanic explosion that scattered ashes as far as north Malaya. This was doubtless the time, at the end of the Pleistocene period, when Jarak was denuded of life and started its career of acquisition as an oceanic island.

In November 1958, we revisited Jarak for some special investigations. At first, everything seemed unchanged. The rats were still swarming everywhere, but we noted three remarkable differences. First and most obvious, the populous land crabs had vanished, which meant that they had gone down to the sea to spawn. In due course, young crabs would swarm over the island again up to the top of the hill. Some observers had vividly described the mass migration of such crabs into the sea or the advancing armies of young ones coming back again: they can be heard from a considerable distance and the whole ground seems to be flowing. I am very sorry never to have had the unpleasant but unforgettable and exciting experience of being lapped by such living waves on Jarak.

The other two extraordinary features were obvious only when we started to trap and dissect the rats. The unusually heavy loads of worm parasites had disappeared—a whole new genus of worm in the liver seemed to have vanished. And the rats had stopped breeding.

Over 160 female rats were examined by Lim Boo Liat of the I.M.R. staff; not only was there not a single pregnancy, but there were not even any uterine scars, showing that breeding had ceased some time before. Over 160 males were also dissected, but of these only a mere half dozen had testes approaching in weight those of a mature male. Of course, it was impracticable to discover if the rats over the whole island were in the same state.

Formerly every rat was infested by the worm *Capillaria hepatica*, many so heavily infested that very little liver substance could be seen amongst the solid yellow masses of *Capillaria* eggs cramming the bile-ducts. Now the infestation was light, many rats apparently quite free from these worms, and not a single one of over 400 was very heavily infested. Whereas formerly a number of rats were infested by a peculiar roundworm in the liver described in a new genus as *Hepatojarakus malayae* by Yeh in 1955,[17] we could no longer find this worm. Nor could we find the little *Nippostrongylus* which used to lie in tiny red coils dotted over the intestinal mucosa just below the surface.

The detailed results of this visit have not yet been published. [18] The findings would be valuable in comparison with later surveys made at intervals, but by themselves they only tell us that the population of rats seemed to be about to 'crash' to some lower level at which breeding would start again. The cause of the disappearance of the worms is open to speculation. Unfortunately, it later became too dangerous to work on this island. Some of my zoological colleagues have had personal experiences of the fantastic peaks in lemming populations in the Arctic and the crashes which follow, but that visit to Jarak is my only personal experience of a large-scale change in the dynamics of a pullulating rat population. It was a novel experience to find the land crabs temporarily absent—the place did seem different without them—but this was not wholly unexpected. It would have been a shock to find that the rats had almost vanished; but the impact was much greater over 2 or 3 days when we realized that something startling was happening to our Jarak rats. I am sometimes subdued at the thought of this afforested land mass rising out of the sea, swarming with rats: perhaps 50 000 of them, almost entirely all adults, with sex at a standstill.

This island has a nostalgic fascination for me, and I sometimes long to be back there on my camp bed with the rats scuffling on the forest floor. But I have also felt nostalgic for many other places, including the asphalt jungle of San Francisco. My thoughts have gone back to Jarak when, from the air over the approach to the airport, I have seen the myriads of scurrying humans, and the crowded beetles jostling along the highways.

No human problem is greater than that of over-multiplication. While we are trying to cope with this, we must see what can be done about the consequences of multiplication that are already upon us.

A broad horizon which has expanded tremendously since the war is the field of comparative animal population dynamics. We now know a great deal about the inbuilt neurosecretory mechanisms which, to a varying degree in different species, are among the processes that control the densities of animal populations in nature.[19, 20] These particular mechanisms are of great interest to mankind because man has a biological inheritance of similar mechanisms that are doubtless the basis for his psychological response to stress, especially stress of social or psychological origin. These human mechanisms and human responses probably cannot be fully comprehended without comparative animal studies.

Furthermore, complex genetic changes take place in fluctuating animal populations[21] and these also are relevant to an understanding of human genetics.

Finally, animals and man have special relationships to space. The social use of space (proxemics) has become a topic for special investigation, in which comparative studies of rodents and other animals give us necessary information.[22, 23]

Science is advancing much more rapidly in the world of molecules than in the world of man. We cannot afford to wait too long to redress this difference because our human problems, including those of mental and social health, are becoming intolerable under the weight of the population avalanche. As science is now progressing, we may learn to land on the moon before we know how to live with each other; we may alter the genetic code of unborn infants and even rear them in bottles, before we know how to provide infants and children at large with the elements that are needed for their future social adaptability and mental health. Questions that we want answered most urgently arise from problems at the level of human society, the complexity of which should no longer fill us with dismay. They do not arise at the level of molecules even though ultimate explanations are partly molecular.

In the broad field of preventive medicine something should be added to epidemiological studies, and that is the comparative investigation of animals; because so frequently it is only by comparing that one may fairly judge.

REFERENCES

1 KUWATA, T., BERGE, T. O. and PHILIP, C. B. (1950). A new species of Japanese larval mite from a new focus of tsutsugamushi disease in southeastern Honshu, Japan. *J. Parasitol.* **36**, 80–3.
2 TAMIYA, T., Editor (1962). *Recent Advances in Studies of Tsutsugamushi Disease in Japan*, 1–308. Tokyo, Medical Culture Inc.
3 SASA, M. (1954). Comparative epidemiology of tsutsugamushi disease in Japan. *Jap. exptl. Med.* **24**, 335–61.
4 SASA, M. (1956). [Tsutsugamushi and tsutsugamushi disease.] Igakushoin. Tokyo, 1–497. (In Japanese.)
5 KUMADA, N. (1959). Epidemiological studies on *Trombicula* (*Leptotrombidium*) *pallida* in Japan, with special references to its geographical distribution, seasonal occurrence, and host-parasite relationships. *Bull. Tokyo med. & dent. Univ.* **6**, 267–91. [See also ASANUMA, K. (1959). Evidences for *Trombicula scutellaris* to be the vector of scrub typhus in Chiba Prefecture, Japan. *Jap. J. san. Zool.* **10**, 232–44.]
6 PAVLOVSKY, E. N., PETRISHCHEVA, P. A., ZASUKHIN, D. N. and OLSUFIEV, N. G., Editors (1955). [*Natural Nidi of Human Diseases and Regional Epidemiology*—proceedings of a joint conference in 1954] (in Russian). Leningrad, Medgiz. Also, PAVLOVSKY, E. N. (1966). *Natural Nidality of Transmissible Diseases with special Reference to the Landscape Epidemiology of Zooanthroponoses*. English translation by F. K. PLOUS, edited by N. D. LEVINE. Urbana and London, University of Illinois Press.
7 PAVLOVSKY, E. N. (1965). The formation of natural foci of diseases in towns and their return from anthropurgic environment to Nature. Introductory paper, pp. 23–37, *in* B. ROSICKÝ and K. HEYBERGER, Editors, *Theoretical Questions of Natural Foci of Diseases*, Proceedings of a Symposium in 1963. Prague, Czechoslovak Academy of Sciences.
8 SWAMY, T. V. and DUTTA, B. B. (1953). Epidemiology of XK typhus in Jamshedpur. *Ind. med. Gaz.* **88**, 522–5.

9 RUDNICK, A. (1966). *Aedes aegypti* and haemorrhagic fever. (Proc. Seminar on the ecology, biology, control and eradication of *Aedes aegypti*). *Bull. World Hlth. Org.*, **36**(4), 528–32.

10 HARRISON, J. L. (1956). The effect of withdrawal of the host on populations of trombiculid mites. *Bull. Raffles Mus., Singapore* **28**, 112–19.

11 PHILIP, C. B. (1965). Scrub typhus and scrub itch. Chap. XI, pp. 275–347, in *Preventive Medicine in World War II*. Vol. VII, *Communicable Diseases. Arthropodborne Diseases Other Than Malaria* (page 289). Washington, D.C., Office of Surgeon General, Department of the Army.

12 AUDY, J. R. (1958). The localization of disease with special reference to the zoonoses. *Trans. roy. Soc. trop. Med. Hyg.* **52**, 308–34.

13 ROSICKÝ, B. and HEYBERGER, K., Editors (1965). *Theoretical Questions of Natural Foci of Diseases*. Proceedings of a Symposium in 1963. Prague, Czechoslovak Academy of Sciences. [Especially Section 3, The influence of man on the existence and development of natural foci of diseases—11 papers, pp. 227–352.]

14 AUDY, J. R. (1950). A visit to Jarak island in the Malacca Straits. *Malay Nat. J.* **5**, 38–46.

15 AUDY, J. R., HARRISON, J. L. and WYATT-SMITH, J. (1950). A survey of Jarak island in the Malacca Straits. *Bull. Raffles Mus., Singapore* **23**, 230–61.

16 TWEEDIE, M. W. F. (1950). A note on Pulau Jarak considered as an oceanic island. *Bull. Raffles Mus., Singapore* **23**, 262.

17 YEH, L. S. (1955). A new bursate nematode *Hepatojarakus malayae* gen. et. sp. nov. from the liver of *Rattus rattus jarak* (Bonhote) on Pulau Jarak, Straits of Malacca. *J. Helminth.* **29**, 44–8.

18 INSTITUTE FOR MEDICAL RESEARCH (1959, 1960). Division of Virus Research and Medical Zoology: Investigations on Jarak island. *Rep. Inst. med. Res. Malaya*, 1958, 68–74; 1959, 64.

19 CHRISTIAN, J. J. (1963). Endocrine adaptive mechanisms and the physiologic regulation of population growth. *In* W. V. MAYER and R. G. VAN GELDER, Editors. *Physiological Mammalogy*. I. *Mammalian Populations*, pp. 189–358. New York and London, Academic Press.

20 MULLEN, D. A. (1965). An alternative to the 'ultimate factor' hypothesis of population control in microtine mammals. *Amer. Zoologist* **5**, 359.

21 CHITTY, D. (1964). Animal numbers and behaviour. *In* J. R. DRUMMOND, Editor. *Fish and Wildlife: A Memorial to W. J. K. Harkness*, pp. 41–53. Toronto, Longmans, Canada.

22 CALHOUN, J. B. (1963). The social use of space. *In* W. V. MAYER and R. G. VAN GELDER, Editors. *Physiological Mammalogy*. I. *Mammalian Populations*, pp. 1–188. New York and London, Academic Press.

23 HALL, T. E. (1966). *The Hidden Dimension*. Chicago, Doubleday.

Appendix 1

CLASSIFICATION OF ACARINA

To put the family Trombiculidae in its place among other acarines, a classification of the Acarina which is to be recommended is that summarized by G. Owen Evans, J. G. Sheals and D. MacFarlane (1961). *The Terrestrial Acari of the British Isles*, vol. 1, *Introduction and Biology*. Chapter 2, pp. 35–60. London, British Museum.

Phylum ARTHROPODA
Subphylum CHELICERATA

Class MEROSTOMATA. King Crabs, marine
Class ARACHNIDA
Includes 16 subclasses of which 5 are extinct. The others include the Scorpions, Pseudoscorpions, Solifuges, Whip-Scorpions, Harvestmen (Opiliones), Spiders (Araneae), and Acari:

Subclass Acari (Acarina)

A. Superorder *Acari-Anactinochaeta*
The name signifies that the setae are not optically refractile (birefringent) as detected by polarized light.
1. Order *Mesostigmata* – Mostly free-living in the soil but some are parasitic on vertebrates and invertebrates ('gamasoid' or 'parasitoid' mites in the families Laelapidae, Dermanyssidae etc. mostly infesting rodents and birds).
2. Order *Metastigmata* – Ticks, the largest acari.

B. Superorder *Acari-Actinochaeta*
The true setae are birefringent due to the presence of a core of actinochitin.
1. Order *Prostigmata* – The most heterogeneous order of mites. Includes some plant-parasites such as the gall-mites, and the superfamily Trombidioidea which includes the family Trombiculidae. The larvae of Trombiculidae are, with very

few exceptions, parasitic on vertebrates, especially small mammals. The larvae of closely related families are parasitic on insects and, to some extent, on reptiles.
2. Order *Astigmata* – Includes the Sarcoptiform mites such as *Sarcoptes scabei* (the itch-mite of scabies), related mange-mites, and the feather-mites of birds.
3. Order *Cryptostigmata* – Includes the oribatids or beetle-mites in the soil.

Appendix 2

SCRUB-ITCH: SOME QUOTATIONS FROM THE LITERATURE, ESPECIALLY OUDEMANS, 1912[1]

WITH reference to the Malaysian zoogeographical region, the following is a translation of comments and a number of interesting quotations by Oudemans himself ('Larven von Thrombidiidae und Erythraidae', *Zoologische Jahrbücher*, Suppl. 14, 45–62, 1912). Although this passage is under the heading of biology of the species *Schoengastia vandersandei* (Ouds., 1905), some or even many of these references are to at least one other species, the *gonone*, *Eutrombicula wichmanni* (Ouds.). Recent investigations (James Gentry, personal communication) have shown that the scrub-itch chigger *Schoengastia schueffneri* (Ouds.), hitherto known from Sumatra, is very common around Kudat in Sabah (North Borneo). It may readily have been distributed east of Wallace's line and some of these old accounts may refer to this species also.

Oudemans, writing in German, quotes the originals extensively in French, German, Dutch, and English, but the following is translated into English. I have not consistently adhered to Oudemans' separation of paragraphs.

Biology: Since New Guinea and the surrounding islands have been visited by Europeans, the visitors have frequently complained about an itch, produced by an invisible insect or a tick, on the legs of those people walking through grass and forest. The information which I have concerning this horrible plague I have obtained from the late Mr. C. A. J. van der Sande, physician to the Netherlands New-Guinea expedition of 1903.

1845: The oldest report of the New Guinea harvest mite was left to us by J. H. de Bondyck Bastiaanse, Captain of the 'Iris'. Occasionally he took a small hike of a few hours along the beach and forest: on his return he always started itching. He also mentions that the itch differs in intensity in different people: unbearable in some, in others it soon decreases. The usual treatment leads to rapid

---

[1] I am grateful to Mrs Rosalie Watkins for her help in translation.

improvement in some individuals, whilst in others, improvement is slow. His work is called: 'Voyages faits dans les Moluques à la Nouvelle Guinée et à Célèbes'; Paris, 1845.—page 19 we read: 'The next day we all felt again a horrible itch, especially on all parts of our legs.' And page 26: 'The evening before it had grown late when we returned on board. This time also, as in our preceding excursion, we felt an intense itch, principally in our legs. Our physician, who had tried several external cures on himself, having suffered from the itch longer than we had, left us to the care of nature.'

1869: The fact that Wallace, in his work 'The Malay Archipelago: the land of the Orang-utan and the bird of Paradise, 1869' did not meet the bush-mite in Dorei, was due to the fact that he did not penetrate the forest.

1875: The well-known investigator C. B. H. von Rosenberg, tells us in a 'Travels to Yellow-finch-bay in New Guinea in the years 1869 and 1870' ['Reistochten naar de Geelvinkbaai op Nieuw-Guinea in de jaren 1869 en 1870'], 's-Gravenhage, 1875, p. 40: 'A mass of plaguing insects turned our stay into hell. Amongst these insects there were colossal spiders (Epeira); ants (Asellus) [sic!— *pissenbedden*] and a sort of ground-mite which was the worst. The mites occurred in between and beneath the floor-boards; by the second day they had crept into the clothes and onto the body; there seemed to be no way of getting rid of them; they produced an unbearable itch.' [These *could* have been mesostigmatic mites from rodents.]

This occurred in February 1869 in Rumsaro in Geelvink Bay (Yellow-finch-bay) where he was forced to land after the ship had been damaged by boulders. The wood of the base of the hut, in which he stayed temporarily, consisted of beams and boards in disorderly arrangement on the damp soil. That the mites attacked his entire body was probably due to the fact that he lay down on this earth to sleep. The itch was for the first time ascribed to an animal: a sort of mite, a ground-mite. We have to accept that v. Rosenberg saw and correctly identified the mite!

[1878: Newspaper accounts by M. Achille Raffray, in charge of a scientific mission for the Minister of Public Instruction, 1876–7, contain no references to scrub-itch.]

1879: According to P. J. B. C. Robide van der Aa: 'Travels to Netherlands New Guinea undertaken under order of the Government of the Dutch East Indies in the years, 1871, 1872, 1875–1876' ['Reizen naar Nederlandsch Nieuw-Guinea ondernomen op last der Regeering van Nederlandsch Indië in de jaren 1871, 1872, 1875–

1876'], 's-Gravenhage, 1879, Mr. J. E. Teysmann made his first scientific visit to New Guinea in 1871 aboard the 'Dassoon'. From Salawatti Teysmann visited several islands in the East-straits, amongst others Rumbobo—in September of that year. These trips never lasted more than one day; he never slept on land. This explains adequately why he only suffered from itching on his legs. The ankles and parts where the edges of the shoes touched, were particularly characteristic. From Teysmann's diary, or from Reports of the Dutch East Indies Government various articles by Robide van der Aa were published. One of these reads as follows: 'After a walk on Rumbobo I noticed, as did all the other men, that I was accompanied by hundreds of red bubbles [*blaasjes*] above the ankles, which produced an intolerable itch. Some believed it was the result of the seawater through which we had to wade to reach the boat; but I said it was some sort of flea, which occurred particularly in the Maleo-nests. Rubbing the area with spirits gave much relief; after 3 days the itch decreased, while the bubbles gradually disappeared completely.' [The *Maleo* is the bush-turkey, *Megapodius*. The Australasian Megapodes or mound-builders make bulky, communal, and repeatedly-used nests of earth and decaying leaves, within which they bury their eggs, warmed by the decaying matter. The young can fly immediately after they emerge. These nests become very heavily infested by *E. wichmanni*, and perhaps also *S. vandersandei*. It is not certain if the 'red bubbles' referred to engorging red chiggers or to haemorrhagic vesicles produced by them in sensitized persons.]

As we see, the accompanying men ascribed the itch to the seawater, whilst Teysmann attributed it to fleas of the Maleo-birds. None of them realized the true cause—the ground or harvest mite!

1880: L. M. D'Albertis. 'New Guinea: What I did and what I saw' in 2 volumes, London. This well-known investigator tells us in volume 1, p. 272 (April 4, 1875), while he was landed on Yule Island opposite British New Guinea: 'For some nights we have not been able to sleep, owing to mosquitoes and sandflies. These small and almost microscopic insects are terrible enemies, and put us to real torture. My people, to defend themselves against their attacks, sleep in an open place, surrounded by great smoky fires.' [This could readily be a reference to sandflies or to such pestiferous flies as *Leptoconops*, and not to mites; but see further.]

Volume 1, p. 283 (April 13, 1875): 'They (the Papuans) usually sleep in hammocks, which they make very well, and under which fires are alight all night, to keep off the cold and damp, and those pests, the mosquitoes and sandflies.' Volume 1, p. 282 (April 14,

1875): 'I did not leave the house to-day, being kept at home by a sore foot.'

Volume 1, p. 404; from Yule Island and Hall Sound in British New Guinea: 'Small insects, such as gnats and sandflies, are much too abundant, and are a continual torment. There are day-gnats and night-gnats, of all colours and of all sizes. The sandflies are our next greatest enemies, and they penetrate through everything. Fortunately, however, they only make an appearance during the first days of a new moon.'

With 'my people', p. 272, he meant the people from Bismarck-Archipelago and from the New Hebrides, also natives: Papuans were also affected. Probably the 'sore foot' of D'Albertis was the result of attack by our mite.

Noteworthy is the assurance of D'Albertis that the 'sandflies' appeared only after each new moon. Since the remarkable life-cycle of 3 species of Palolo worms is known, all reports of animals and plants showing sensitivity to moon-phases should be noted very carefully, and whenever possible, investigated, and not discarded as old-wives' tales.

Volume 2, p. 34: 'For some days past all on board have complained of a violent itching in different parts of the body, without finding out any cause for it, but to-day an almost microscopical insect was discovered, which has either attacked us in the forests, or has been carried on board in the skin of the gowras and other birds, which mostly live on land. I found that certain red lumps on the skin of these birds are actually formed of hundreds of these almost invisible creatures. The root of each feather becomes a pleasant abode for them. They must also possess immense productiveness: in fact our bodies were entirely covered with them. It was useless to wash ourselves, either with hot water and soap, or acid, or eau-de-Cologne; but to-day we had recourse to a new expedient, that of washing ourselves with petroleum. At Yule Island I found this process efficacious in similar cases, and also that the ulcer produced by the disease called cascado dies away after repeated applications of the oil. I also tried it at Andai in 1872, and at Yule Island this year.'

Here it should be noted that, when the *Thrombidium* larvae have attached [Oudemans is here speaking of *Eutrombicula wichmanni*], they will not let go, thus there is no possibility of their having been conveyed from forest to boat by means of crown-doves (?) [*Goura*]. [Here Oudemans may be mistaken, for if some species of chiggers have very recently attached, or have started attaching, and the host is killed, they will detach themselves and at least a proportion will attack a new host. Several of us tested this during the war.] Obviously persons only start itching several hours after being attacked by mites in the forest. The bugs are 6-legged larvae, which cannot reproduce; after a filling bloodmeal [actually tissue-juice] they transform to 8-legged nymphs, which, hatching from the larval pelt, drop to the ground and become predaceous, sucking dry all smaller, and also larger creatures. [We do not know exactly what evidence Oudemans had for this but his statement is correct. Later workers got the idea that the post-larval stages were vegetarian, a most unfortunate misconception referred to in Chapter 4, page 90.] With their maxillary palps they attach to the prey; with their mandibles they cut these up, allowing the juices to flow out, which are then sucked up by the canaliculate lower lip (overgrown maxillary bases). These mites taken from the *Goura* heads which I have examined, were not *Schoengastia vandersandei* but *Microthrombidium* [*Eutrombicula*] *wichmanni*. The possibility still remains, however, that the first mentioned species may also feed on man in New Guinea.

In any case D'Albertis definitely saw the mites. Notwithstanding the fact that he called them 'sandflies' he described them as 'almost microscopic insects', said that they 'penetrate everything', and regarded them as identical with the parasites of the crown-doves, which were 'red almost invisible creatures'.

1884: A. Haga. 'Netherlands New Guinea and the Papuan Islands. Historical reports.' [Nederlandsch Nieuw Guinea en de Papoesche Eilanden. Historische Bijdrage.'] ± 1500–1883. Batavia and 's-Gravenhage, 1884. Haga described the work of the people aboard the 'Triton' and the 'Iris' in the Merkus area in Triton Bay, southwest coast of New Guinea, 1828. The sailors had to clear an area to build the Fort Dubus. Haga tells us, volume 2, p. 25: 'Felling the thick tree-growth was a very difficult task, which was, moreover, made more difficult by the presence of myriads of bloodsuckers, and the appearance of an unbearable itch, the cause of which we could not find.'

Haga did not mention the source from which he obtained this. In a footnote he says: 'In the following tale, I have also used the travelogue of Modera.' However, if we check Modera's work, the title reads: 'Tale of a voyage to and along the Southwest coast of New Guinea, in 1828 by Z. M. corvette Triton, and Z. M. colonial schooner Iris', Haarlem, 1830, we can find the page referring to the building of the Fort Dubus (p. 96—which on 6 July 1928, the sailors started by felling the trees; and, pp. 133–8, that many of the men suffered from marsh-fever and died) but there is no mention of itching.

APPENDIX 2

Dr. Salomon Muller. 'Travels and Researches in the Indian Archipelago carried out on orders of the Netherlands Indian Government between the years 1828 and 1836.' ['Reizen en Onderzoekingen in den Indischen Archipelago gedaan op last der Nederlandsch Indische Regeering tusschen de jaren 1828 an 1836', Earste Deel, Amsterdam, 1857.'] Part 1, Amsterdam, 1857, is not cited by Haga, although he was on this voyage. Muller does not mention anything about the crew suffering from itching whilst constructing Fort Dubus. Haga, therefore, probably obtained his report from the Archive in Batavia, e.g., from the Report of Pierson, ship's physician to the expedition, 1828.

1885: James Chalmers and Wyatt Gill do not mention any mite plague. Their work is called: 'Work and adventure in New Guinea 1877 to 1885: with two maps and many illustrations from original sketches and photographs.' London, 1885. The part written by Gill bears the title: 'Seven weeks in New Guinea', London, 1885.

1888: Nor do we find any mention of mite disease in Rev. S. McFarlane's 'Among the Cannibals of New Guinea', London, 1888.

1888: We also search fruitlessly in John Strachan: 'Explorations and adventures in New Guinea', London, 1888. He was ship's captain and buyer.

1891: In 'Two years among the savages of New Guinea, with introductory notes on North Queensland', by W. D. Pitcairn, London, 1891, there is also no mention about itching or mites.

1891: After von Rosenberg in 1869 (see above 1875) and D'Albertis in 1875 (see above 1880) Lauterbach also saw the mites, but did not recognize them as ground-mites!—in the 'Reports on Kaiser-Wilhelms-Land and the Bismarck Archipelago'. Distributed by the New Guinea Company at Berlin, 1891, Lauterbach published his report on his travels: 'An expedition to survey the hinterland of Astrolabe-Bay.' ['Eine Expedition zur Erforschung des Hinterlandes der Astrolabe-Bai.'] On p. 38 we read:

'To the unpleasant additions belong land-leeches which, however, appeared singly, and especially the so-called bushlice [*Buschläuse*]. Here I would like to say a few words about this great plague of New Guinea forests. The bushlouse belongs to the mites and is probably closely related to our ticks. It is microscopically small, and only on very close inspection can they be seen as tiny red points on the skin. The animal, although occurring in the bush, prefers open spaces. Here they were particularly numerous amongst the reeds. The animals are brushed off by movements of the plant; they also creep up from the soil. They immediately start burrowing into the skin, and produce intense itching. Very soon a localized inflammation follows, and if the animal is not removed, the constant scratching results in wounds which rapidly start festering. Particularly my Malays suffered very much from this. The best protective measure against this evil is wearing tight boots of sailcloth and bathing immediately after a longish stay in the forests, together with a change of clothes. The clothes must be washed immediately since the animals remain alive for long periods in these. Carbolic acid (3%) is good for wounds already present, and helps against the itch.'

We see that Lauterbach connected the *Thrombidium* larvae, which he correctly described as microscopic red dots, with the ticks Ixodidae, which is by no means true! Yes, broadly speaking, all mites are related: but strictly speaking the Ixodidae form a group known as the Metastigmata, whilst the Thrombidiidae, together with numerous other families, belong to the Prostigmata.

1892: In the work: 'British New Guinea', by J. P. Thomson, F.R.S., G.S., with map, numerous illustrations and appendix, London, George Philip and Son, 1892, the mite plague is not mentioned.

1896: The following report deals with the fate of an expedition, whose aim was to travel straight through the wilderness from the mouth of Franziska River in Bayern-Bay, on the east coast of German New Guinea, to Motu-Motu, on the south coast of British New Guinea. The expedition started August 1895, led by Ehlers, and accompanied by police subordinate officer Piering and 21 bearers. Both were killed by the bearers; of the latter the following report came back (in 'Reports on Kaiser-Wilhelms-Land and the Bismarck Archipelago'. Distributed by the New Guinea Company in Berlin, 1896, p. 52):

'The landleeches and bushticks [*Buschzecken*] rapidly became a great nuisance, since they produced deep, strongly bleeding wounds, which would not heal. The two whites fared no better than the blacks, since their clothes were gradually ripped off by the thick bush, and they were then attacked by these bloodthirsty insects in the same way as the bearers.'

Naturally one replaces 'bushticks' with 'bushmites' or better 'ground mites'. [But these do not draw blood.]

1898: As opposed to this, we find no mention of mites in H. Caylay Webster's 'Through New Guinea and the cannibal countries', with illustrations and map, London, F. Fischer Unwin, 1898.

1899: After the findings of von Rosenberg (see above 1875), D'Albertis (1880), and Lauterbach (1891), the doubts of Hagen sound strange. He appears convinced, that completely different causes—which really are present during a trip through primaeval forests—result in these known symptoms. In his work 'Amongst the Papuas. Observations and studies of the Land and People, Animal- and Plant-kingdoms in Kaiser-Wilhelms-Land' [Unter den Papuas. Beobachtung und Studien über Land und Leute, Thier- und Planzenwelt in Kaiser-Wilhelmsland], 1899, p. 72:

'One of our bearers complained about "bushgnats" [*Buschmucker*]. His ankles were covered with small, inflamed, painful nodules, in the same way as we would be after insect- or thorn-pricks or such-like experiences. In the forest with thin clothing this is very possible; and since the path goes now through water, now through slime and mud, it is understandable that these small wounds become inflamed and result in large, obstinate and long lasting festering sores on the ankles, which everyone who has lived in the tropics knows. Here in New Guinea, however, one is not satisfied with this explanation; here the mythical "bushgnats" are responsible—I would like to know who invented this name! Supposedly there are small red mites which bore into the skin and produce the inflammation. I frequently saw the usual symptoms occurring with acclimatization during the first year in the tropics such as festering sores and inflammation, also in Sumatra, where the "bushgnats" are not known; but no one has yet been able to show me the mite. Also our patients have had no luck in this direction; yet the possibility cannot be ignored.'

1899: In the 'Annual Report on British New Guinea, July 1, 1897 to June 30, 1899', appearing in Victoria, 1899 (and published by the government), p. 8, we read: 'we all suffered greatly with shrub [*sic*] itch'.

1901: In the same report, 2 years later, July 1, 1899 to June 30, 1900, published in Victoria, 1901, we find again in the description of a trip, p. 34: 'we were much tormented by shrub [*sic*] itch'.

1902: The honourable Mr. G. A. J. van der Sande, who accepted the medical responsibility for the following expedition, asked his colleague and friend Mr. van der Willigen, to acquaint him with the diseases there present, and the best methods for curing the diseases. The latter had for several months been ship's physician on the flotilla-leader 'Ceram', which sailed in the vicinity of Humboldt Bay from July to October 1901. The part, dated 1902, that interests us of his reply to the question about the disease, is as follows:

'Another nuisance on small trips inland was the itch, which without exception got everyone on the feet and lower leg areas. Everyone in inland New Guinea complained about this. I do not know which creature [here he uses the Malay word *binatang*, applicable to any living creature] we are dealing with—with regard to the bite. Neither ants, nor worms have I discovered in the skin. Glycerine-chinine which is commonly used against insect bites, was frequently rubbed into the skin, without success. Various ointments were tried, yet the greatest success was still with "cajuput oil" ' [Malay *kayu*, wood; *puteh*, white; producing an eucalyptus-like oil].

We note that van der Willigen could not find the true cause.

1903: Prof. C. E. A. Wichmann, the learned investigator of the Netherlands New Guinea Expedition of 1903, wrote in his '5th Bulletin of the New Guinea Expedition', p. 20 (which appeared July 1903, in: '45. Bulletin der Maatschappij ter Bevordering van het Natuurkundig Onderzoek der Nederlandsche Koloniën'):

Humboldt Bay, May 28, 1903: 'In the Mosso [?] the "bushticks" [*boschteeken*] again produced large wounds, especially on the feet and legs, and I had a bad time. The hardly visible beasties creep under the skin and there lay their eggs, which produce an unbearable itch. Later, when these have developed to larvae, large yellow blisters are formed, sometimes as large as a half-guilder. I had 20 such blisters, all of which were cut open. As a result of this I cannot walk, and our next trip has had to be postponed for several days, especially since half of coolies have malaria, and we have not enough bearers therefore to carry our baggage.' [The reference to the laying and hatching of eggs under the skin is an imaginative excursion unexpected in a scientist such as Wichmann. He was perhaps thinking of scabies mites and simply assumed that scrub-itch must be scabies-like in its etiology. As Oudemans notes, below, Wichmann later changed his mind.]—and p. 23: 'Unfortunately the ticks (sandfleas) [*zandvlooien*] have again attacked us badly, and my joints were covered with new boils and blisters.'

As we see, Dr. Wichmann called the *Thrombidium* larvae 'ticks' or 'sandfleas'! And he assured us (on which grounds one might well enquire) that they bore into the skin and there lay their eggs!! In a following report the learned gentleman had changed his mind. He says in his '6th Bulletin' (published in September 1903 in the '46. Bulletin van der Maatschappij'):

'July 25, 1903, on board the S.S. Zeemeeuw, near Manokwari Mr. Lorentz and I were, as a result of attack by bushmites, unable to take part in proceedings, so that H. H. van Nouhuys and Dumas on July 8, had to go out alone.'

1904: It is noteworthy that van Dissel did not meet the mites at all. He published his experiences in a report on a 'Land-journey from Fakfak to Sekar', in the utter wilds of New Guinea in August and September 1902 (in: 'Indische Gids', 1904, p. 970). 'Of plaguing animals, apart from the land-leeches [*patjets*, Malay for the common

Haemadipsa zeylanica (Moq.-Tand.)] we had no trouble. Fortunately the small bushlouse, which is so abundant in other parts of New Guinea, and which causes such an unpleasant itch when in contact with the skin, does not occur here. As remedy is used the sap pressed from the leaves of the wild *Piper betle* [sirih].'

In the descriptions of the two following expeditions (in: 'Tijdschrift van het Koninklijk Nederlandsch Aardrijkskundig Genootschap,' Jahrgang 1904, (2), vol. 21, pp. 617 and 789), van Dissel does not mention the matter again, so that one may well conclude that he had no trouble with mites.

1904: W. L. Jens: 'The Papuans of Yellow-finch-Bay'—a presentation with slides given at the Meeting on May 14, 1904, and published in the 'Handelingen van de Nederlandsch Anthropologische Vereeniging', 1904, pp. 45–61. Jens does not appear to know the mites or the mite-plague, at least he makes no mention of it.

1904: In the Supplement to the 'Illustrated London News', October 1, No. 3415, 1904, A. E. Pratt published a report entitled: 'Two years among cannibals. Being some account of the aborigines of Papua, New Guinea and of travel and adventure on that Island'. The author, who was particularly interested in accumulating subject matter for his natural history study, tells us about plants and leaves which produce an intense itch and result in blisters, but about the mites, which might be the cause of the itch, he says not a word.

1905: Mr. G. A. J. van der Sande, physician to the Netherlands Expedition of 1903, on his return to his homeland, asked Professor A. C. Haddon (leader of the Cambridge Anthropological Expedition to Torres Straits) in a letter whether he had not come across mite-plagues in any areas. Prof. Haddon immediately replied that, although his expedition of 1898 had not noticed any mites, Seligmann who had just returned home from British New Guinea, had studied diseases there, and he (Prof. Haddon) had sent van der Sande's letter to Seligmann. Seligmann was physician to the 1898 expedition of Prof. Haddon, and later given to the Daniels Expedition to Woodlark Archipelago. In 1905 he told Mr. van der Sande as follows about the disease:

'I just returned from New Guinea where I was in medical charge of the Daniels Expedition. The symptoms you mention which I always attributed to a tick, though I collected none, are common after moving through the bush in British New Guinea, where the popular term for the lesions is shrub itch.—The tick does not, however, seem to be of universal distribution at least at the same time of the year. Besides suffering from it in British New Guinea proper, I can testify to its prevalence and virulence on Marua (Woodlark Island) off the S.E. extremity of British New Guinea.' [Probably both *E. wichmanni* and *S. vandersandei*.]

Seligmann's observation, that the mites only occur during certain seasons, is probably correct, since in Europe the harvest mite is also known only during a few months. However, the possibility remains that in tropical areas the larvae may be present throughout the year.

1905: Prof. A. Wichmann told me, September 14, 1905: 'Mr. K. M. van Weel, First Office of H.M.S. Zeemeeuw, wrote to me a few weeks ago: 'The animal kingdom (of the Groot-Obi Island) was poorly represented ... only the impossible small fleas, which produced so many wounds in the expedition, were well represented.' The learned gentleman also told me:

'Mr. J. W. van Nouhuys did not see any bushmites on the Soela Islands.' Later, of his own expedition: 'We never met the animalcules in open grassy ground, but always in the shrub, especially where Selaginella was growing. The Assistant resident van Oosterzee also believes this to be the case.'

1906: In 'Two years among New Guinea cannibals', by A. E. Pratt, London, 1906, we read, p. 93: '... the abominable attack of the shrub-itch, a nasty little parasite that the wayfarer brushes from the low herbage as he moves along. This hateful microscopic creature, which is of a bright red colour, gets under the skin and causes terrible irritation. The affection spreads, and if one is so unwise as to scratch the place, there is no hope of relief for at least three weeks. The only satisfactory remedy is to bathe the parts in warm salt and water.'

It is unnecessary to add here that these long-legged larvae cannot burrow into the skin at all but that the skin rises around the larva in a ring-shaped area, enclosing it completely and even growing over the animalcule (of course, with a microscopic hole in the middle), as is the case with *Ixodes* and *Dermatophilus*. See in *Entomol. Ber.* **2**, 56–9.

Mr. van der Sande, whom we have so frequently mentioned, and who accompanied the expedition of 1903, wrote to me in approximately February 1905: 'Noteworthy is the fact that van Dissel denies the presence of mites in the western area, near Fakfak; also that people in the Central-Office of the London Missionary Society, 16 New Bridge Street, London E.C. have never heard "of the existence of such insect troubles".' Mr. P. Wardlan Thompson, who questioned me on this topic in the name of his company, had himself visited all the Mission stations previously. Prof. Alfred C. Haddon and Dr. W. H. R. Rivers, both from Cambridge, who

APPENDIX 2

accompanied the 1898 expedition to Torres Straits and then visited British New Guinea, wrote to me that they had not met or heard of the mites in the areas where they had been. The positive experiences of Dr. Seligmann, who accompanied the 1898 expedition, therefore probably does not date from this year, but from the Daniels Expedition, from which he has just returned home. This is the only positive record which I have with regard to British New Guinea. The English Missionaries and authors don't mention it at all.

The Netherlands marine officers, who went on an expedition on the Meranke river, not far from the Netherlands-British boundaries, made no inland expeditions, and slept along the river bank; they told me they were much troubled by land-leeches, but never by mites. Thus, for instance, the missionary van Hasselt told me that in Dorei the mites did not occur in the houses, but occurred exclusively in long grass. Lauterbach also spoke about 'reeds'. Where this type of grass does not occur, or when the visitor avoids going through it, one is generally not plagued by the mites. It is a fact that the members of our party have more than once noticed the symptoms produced by the mites on their hands, but never on their faces, although these often come into contact with twigs and leaves of the shrubs.

Wallace, who stayed in Dorei for 4 months, only spoke about ants and flies as plagues, never of mites.

The missionary W. L. Jens, at present representative of the Utrecht-Mission Company, Jansberkhof 18, Utrecht, wrote that in Dorei the mites truly did not frequently occur, and that one saw them even less frequently on the Mansinam Island, but that they were very numerous along the coastal areas with black sand, as in Andai, lying slightly south. In the Numfore dialect, the mites were known as 'arkan'.

Our expedition came across the bush-mites on all their trips, also in the country of the Manikion tribes, which lies immediately north of Machner Bay, thus not far from the region in which van Dissel travelled.

Due to the tiny sizes of mites present, there was much difficulty in finding these little animals, and this explained why the various thorough gentlemen, amongst whom were zoologists, could not come to the correct conclusions regarding the true nature and biology of the animalcules. With regard to their biology, Lauterbach says that the mites burrow into the skin; the cousins Sarasin that they burrow into the skin-pores, and may even die there; Prof. Wichmann that they burrow into the skin and lay their eggs there, and that their larvae produce blisters.

If I must add my own experiences to this, I am forced to surmise that the mites remain on the surface of the skin and only stick their mouthparts or mandibles into the skin. This I conclude from the fact that these mites, whether I collected them shortly or some time after the beginning of a trip, I could remove them with a small knife, by carrying this along the surface of the skin; only their mouthparts were damaged. Thus the body of the mite must always remain above the surface of the skin. I have never seen them hanging to hairs, or attached to the openings of the sweatpores.

The symptoms which the mites produce in humans, are, I suspect, the results purely of poisonous saliva, which flows into the wounds. In one person these may be more or less serious, the result presumably, of the amount of poison, and therefore dependent on the length of time the mite remains attached to the skin. Various persons react very differently to this agent. One of the 6 Europeans whom I had under observation felt only a slight itching, which seldom gave rise to scratching, and objectively only slightly red and slightly raised points of small diameter (2-3 mm diameter) could be seen. Only on this person could I find the mites, since they had not been removed by scratching.

Subjective as well as objective symptoms could increase in severity, without there existing a parallelism. The slight itching could become an unbearable itching, so that the patient would scratch himself raw, by which means the mites were naturally removed, and whereupon the itching disappeared in large part.

The pustules could change to boils and blisters, completely filled with a serous fluid, and attain a diameter of 2-3 cm and a height of 8 mm. I have never noticed the mites in these bumps, nor in the wounds themselves, after removal of the serous fluid. In one case I saw extremely large boils, yet the patient had no itching. However when intense itching occurs, no one can withstand the inclination to scratch, thus of course removing the boils. These wounds are characterized by the fact that, due to the poison, they will initially not heal: in fact the exact opposite—the skin swells, and along the edges a band of increasing red and infiltrating area occurs. After the worst stage regeneration of the skin occurs from the outside to the centre; and on the legs particularly of people who are not so young it is characterized by increased pigmentation. In patients who show the most meaningful objective symptoms, the process takes approximately 3 weeks from the beginning until healing occurs. Probably there are greater differences than I have noted amongst the 6 Europeans whom I have observed. An Indo-European was only slightly disturbed by the bush-mites; and as regards our

Malay coolies, at most they had small pustules, which, however, they still scratched to pieces. I never noticed any reaction in the Papuans to the mites.

As a good protective measure for the legs we used oversocks of dense cotton, which were worn over shoes and leg-clothes, and tied above the knees. A sample of these socks was given to us in Makassar by the gentlemen Sarasin. To increase the certainty, the oversocks were initially steeped in cajuput oil; they protected the legs even against land-leeches. However, the oversocks hindered progress considerably, especially in damp spots, and thus were seldom used.

Another relatively good protective measure consisted of equal parts of Peruvian balsam and spiritus fortior—rubbing this on the ankles and feet every morning, seemed to protect these parts fairly well.

Instead of this mixture we sometimes used Unguentum hydrargyri or cajuput oil. The latter showed less reliable results, especially when we were marching through small rivers and morasses—it was washed off more rapidly than Balsamum peruvianum or Unguentum hydrargyri. Both latter methods were generally washed off in a small river after a day's march; we generally used carbol-soap, which eventually definitely killed any mites which had penetrated. A few pustules, which were occasionally noticed, were probably obtained during night-bivouacs. Washing with carbol-soap had no protective action of any duration.

If protective measures were not used, and pustules appeared, the itch rapidly disappeared after immediate rubbing with Unguentum hydrargyri; also damp bandages soaked in 2–3% phenol-acid produced the same results. The sooner these measures were taken the better, so that in any case the disease remained as small pustules; no single area showed inflammation, and in 2–3 days everything was healed. Presumably the mites in the area were killed, before the poison which flowed into the patient became too great. Very frequently these measures were too late, if protective measures had not been taken, in which case one tries to kill the mites after a day's march by rubbing in soap. This was very evident after a visit to the Mios Kairu island: several members of the expedition set out with healthy legs, after a few hours they returned, and were already complaining about the itch. Those who waited until evening without using Unguentum hydrargyri, could already see the appearance of small pustules and later blisters and boils. It is not necessary here to mention that generally immediately after a trip we changed our clothes, and that we did not lay our field-mattresses directly on the soil, but on a piece of waxed taffeta or some other non-penetrable material, in which everything was rolled up during the day. The clothes worn during the day, also socks and sometimes even shoes, were then generally hung in the smoke over a fire to dry. The one or the other way possibly explains why we were only worried on the hands, legs and feet by the mites.

During the approximately 7 months' stay of our expedition in New Guinea none of the members of the expedition could develop an immunity to the poison of the mites; reaction at the end of this period was as severe as in the beginning.

The mites which I collected, were removed by means of a small knife from the human skin, and preserved in 76% alcohol.

The following is a description of *Neotrombicula autumnalis*, the European pest-chigger, from Henry Baker (1753) *Employment for the microscope. In Two Parts. II An Account of various Animalcules never before described, and of many other Microscopical Discoveries.* Pages 393–5. R. Dodsley, London. I am grateful to my colleague Dr P. H. Vercammen-Grandjean, for drawing my attention to Baker's reference. Gilbert White evidently used Baker's account, published nearly 20 years before his letter on the harvest-mite. Baker's fig. 11 does not show the 'Proboscis near 2/3 of their own Length', which is probably a reference to the stylostome or drinking-straw formed in the host's tissues by the mite's saliva: it can be seen attached only to chiggers forcibly detached after they have fed for some time. The 'Antennae' are the palps.

I shall now describe an Insect not found in Water, and very common in Time of Harvest, but of which I have never seen any Drawing or Account. It is called the *Harvest-Bug*: is of a bright red Colour: so very small as to be imperceptible to the naked Eye, and on the Point of a fine Needle resembles a Drop of Blood. A Drawing taken from the Insect preserved in a Slider, and greatly magnified, is given *fig. 11*.

I had often heard of these Insects, but did not give intire Credit to what I heard, 'till a Lady taking this out of her Neck convinced me of their Existence and Taste. They are extremely troublesome to those that walk in the Fields in Time of Harvest, especially to the Ladies, for they know what Skins are finest and easiest to pierce. They have at the Head a Proboscis near 2/3 of their own Length; by which they first make way through the Skin, and then bury themselves under it, (leaving no mark but a small red Spot) and by their sucking the Blood create a violent Itching; a good Remedy for which is a little Hungary Water; though perhaps Spirit of Wine with Camphire might be more destructive to these little troublesome Attendants of Summer Walks. They are I believe frequently carried in the Winds at their Season, for I have since known them attack Ladies in a Garden, which was defended from a Corn Field by a Wall, too high for these Insects to get over any other Way.

They have three Legs on each Side, with four Joynts set with Hair, as the Body is all round. The first pair of Legs arise from the Back, just below the Eyes; the other two pair from the Belly: it has also two short Antennae, one from each Side of the Head, which appears with a Division in the Middle. I have sometimes suspected this little Creature might be a young Sheep-Tick, from its Figure and Way of burying itself: but then it should be found rather where Sheep feed than in Fields of Corn, growing, and before Sheep are suffered to come into those Fields: and it is never got as I have heard in Grass Fields, unless bordering upon Corn; but amongst Wheat it never fails. If any one has a mind to make Trial upon this Insect, how it comes to be amongst Corn only, and yet lives by sucking of Blood, he may easily find Abundance of them: for though they prefer the Ladies, yet they are so voracious, that they will certainly lay hold of any Man's Legs that comes in their Way.

Otto Kepka makes the following references to *N. autumnalis* in a pamphlet on applied parasitology and pest control ['Die Herbstmilbe (*Neotrombicula autumnalis*). Merkblatt Nr. 12 über angewandte Parasitenkunde und Schädlingsbekämpfung.' *Angewandte Parasitologie* **4**, 1–13, 1965].

P. 10, para. 2: Foci of trombidioses therefore develop primarily on cultivated meadows, farmland, in gardens, on the margins of forests . . . on sunny sites where people camp . . . This explains why certain people contract trombidioses—farmers, gardeners, tourists, forest workers, hunters . . .

P. 11, para. 1: In domestic animals and cattle trombidioses has been observed frequently. Considerable damage occurs in chicken farming when late-hatching chicks can be kept outdoors and can be heavily infected. In this case they stop eating and die within a few days from starvation and exhaustion.

P. 11, para. 2: . . . Grown chickens are not immune either, but survive. In turkeys, the immature ones are the most susceptible. Hunting dogs may be attacked massively—nose, eyes, inner side of legs, between the toes, on hips and abdomen ... Horses become irritated and restless from the itching ... cattle, sheep, goats, and cats are attacked in places where there is little hair: lips, cheeks, ears, eyebrows and lids, neck, soles of feet ... Sheep on alpine meadows are often attacked massively from mid-October through November (though by other species of trombiculids) ...

Appendix 3

PEST-ARTHROPODS, WITH SPECIAL REFERENCE TO PEST-TROMBICULIDS (CHIGGERS)

PEST-ARTHROPODS

Arthropods affect animals and man in three main ways: (a) by simply bothering them so that they are unable to feed or sleep satisfactorily; (b) by causing excessive loss of blood, well recognized in the case of tick-anaemia, although blood-sucking flies may do the same; and (c) causing disease, by crippling the animals (as the chiggers at the fetlocks of horses), or by inoculating noxious elements in their saliva (e.g., tick-paralysis, systemic allergic reactions), or by invading the body or its cavities (e.g., warbles, bots), or by transmitting infections. Pest-arthropods are those that annoy by painful bites, by causing allergic reactions, by being attracted to the eyes, nose, and mouth, or by imbibing excessive amounts of blood.

DIPTERA AND BEHAVIOUR OF UNGULATES AND MAN

Biting flies—especially tabanids (deerflies, horseflies), culicids (mosquitoes), and simuliids (blackflies)—may so trouble deer, caribou, and other ungulates that their movements and perhaps even numbers and breeding are affected. Darling (1937) counted an average of 30 bites a minute on face and feet of red deer in Scotland by a vicious fly *Haematopota pluvialis*. The deer are forced up into high ground as these flies start appearing in the lowlands. Harper (1955) describes the effects of hordes of mosquitoes (*Aedes*), blackflies (*Simulium*), warble flies (*Oedemagena*), and nostril-flies (*Cephenemyia*) on reindeer in North America. 'Is it any wonder, then,' he asks, that the Barren Ground Caribou 'hasten throughout August toward or into the shelter of the [fly-free haven of the] woods, to gain freedom from the winged scourges of the Barrens?' *Cephenemyia nasalis* larvae may number a thousand in a single reindeer in the tundra of arctic Russia, and this and other flies are important factors in the ecology of these animals (Sdobnikov, 1935). Schurz (1961: 179–83) gives a vivid account of pest-arthropods in Brazil. Richard Garcia and Frank Radovsky have described to me how horses in California will, during the deerfly season, seek shelter in pairs, head-to-tail under trees so that each can swish the head and neck of the other; or they push their heads into dense foliage. Philip (1931) gives a vivid description of tabanids attacking livestock in Minnesota. He estimated a minimum loss of 300 cc of blood by an animal on a sunny July day.

The most notorious example of control of animal populations by an arthropod is the transmission of trypanosomes by tsetse flies (*Glossina*) in Africa. Since the grazing of large herds of cattle, deer, and antelope has a marked effect on the vegetation, the depredations of the tsetse fly may greatly influence the vegetation indirectly and perhaps set complex cycles in motion (e.g., Thomas 1943).

DARLING, F. F. (1937). *A Herd of Red Deer; A Study in Animal Behaviour.* London, Oxford University Press.
PHILIP, C. B. (1931). The Tabanidae (horseflies) of Minnesota, with special reference to their biology and taxonomy. *Minn. tech. Bull.* **80**, 1–128.
SCHURZ, W. L. (1961). Brazil, *The Infinite Country.* New York, E. P. Dutton.
SDOBNIKOV, V. M. (1935). [Relationship of the Reindeer with the animal world of the tundra and of the forest] (in Russian). *Otd. Ottisk* Sep. edit. pp. 5–66. [Quoted by BOURLIERE, F. (1956). *The Natural History of Mammals.* New York, Alfred Knopf.]
THOMAS, A. S. (1943). The vegetation of the Karamojo District, Uganda. *J. Ecol.* **31**, 148–78.

TICKS AND BEHAVIOUR OF ANIMALS AND MAN

The spinose ear-tick, *Otobius megnini* has actually adapted its life-cycle so that the host's attempt to relieve its suffering, by rubbing its ears against trees or posts, help the transmission of the tick from host to host. An animal that simply frequents the same shelter for long enough, perhaps by successive generations, may allow such heavy infestations of some parasites to build up that it must either succumb or be driven away. An example may possibly be found among porcupines and soft-ticks (*Ornithodorus*). Man may do the same and it would be profitable for the historian and the anthropologist to search for this.

In connexion with the punishment of culprits by ticks mentioned in Chapter 1, Petrishcheva has published a delightful and informative collection of anecdotes describing fieldwork. She described the '... confinement of culprits in cells with ticks as a form of slow torture in the Khiva Khanate. The cells were known as "bug-ridden" ... ' She first learned of this by listening to the song of a *bakhshi* or bard, with its sad ending when the hero was thrown to the ticks. Repeated attacks by soft-ticks can become intolerable to man and Walton (1964) has described some of the human responses to this. Some general aspects of man-made maladies of this sort have been summarized by Audy (1965).

AUDY, J. R. (1965). Types of human influence on natural foci of disease. *In* B. ROSICKÝ and K. HEYBERGER, Editors. *Theoretical Questions of Natural Foci of Diseases.* pp. 245–51. Proc. Symposium, Prague, 26–29 November 1963. Prague, Publishing House of the Czechoslovak Acad. Sci.
PETRISHCHEVA, P. (1960 – translation undated). *Forewarned is Forearmed.* Moscow, Foreign Languages Publishing House.
WALTON, G. A. (1964). The *Ornithodorus moubata* superspecies problem in relation to human relapsing fever epidemiology. *Symposia Zool. Lond.* 6, 93–156.

ACARIOSIS OR INFESTATION BY MITES

There is incidentally much inconsistency in the use of the endings -*osis* and -*iasis* in parasitology. A logical system favoured by Norman Levine and many others is to use -*iasis* simply to imply infestation (i.e. without disease, a common occurrence) and -*osis* for infestation with derangement or pathology. This is logical and consistent, improves precision of speech, and draws attention to the frequent occurrence of potentially pathogenic organisms without disease. But it will surely meet with much opposition.

Acariosis or, perhaps better, *acarine dermatosis*, is the name given to attacks by acarines, especially if they are multiple. Trombidiosis (-*iasis*) or scrub-itch is a special case. Attacks by acarines other than chiggers (e.g., larval ticks, mesostigmatic bird-mites) may be confused with scrub-itch but many such acarines will invade houses when their habitats outside (caves, attic, garden) are disturbed or their hosts depart. Chiggers apparently never do this (although they may be brought in with cut grass). Furthermore, many such acarines give some irritation immediately after they bite, quite apart from the delayed irritation of the developing allergic reaction. Chigger bites are not immediately noted, nor are the chiggers felt tickling the skin as they move.

Acarioses may be caused by at least the following groups of acarines:

Trombiculidae: Trombidiosis or Trombidiasis ('chigger-iasis' of at least one author), discussed in more detail below.

Ixodidae: Larval hard-ticks may attack in large numbers. Soft-ticks (*Ornithodorus*) may frequent huts, caves, hollow trees, and animal burrows.

Sarcoptidae: *Sarcoptes*, *Notoedres* etc., scabies and mange-mites which pass their whole life-cycle immediately under the skin. Some related mites in stored products cause occupational dermatoses—vanillism (vanilla pods), copra-itch, grocer's itch —and some of these mites or their relatives may infest mattresses.

Dermanyssidae and Laelapidae: Bird-mites from nests in caves, chicken-houses, pigeon-lofts; rodent mites from rats and mice. These, in particular, are associated with psychological disturbances (acarophobia) partly due to morbid fear of verminous attacks but most often due to misdiagnosis and mistaken 'treatment' for psychogenic dermatoses. Sambon (1928. *Ann. trop. Med. Parasit.* **22**, 67–132) gives a vivid account of bird-mite infestation of man.

Tetranychidae: These predatory mites on plants may cause quite sharp bites, but usually without an allergic response.

Erythraeidae: Newell describes the invasion of a children's hospital ward by mites which are normally pollen-feeders. The mites were contaminated with staphylococci acquired after entry into the ward. (Newell, I. M. (1963). Feeding habits in the genus *Belaustium* (Acarina, Erythraeidae), with special reference to attacks on man. *J. Parasitol.* **49**, 498–502.)

APPENDIX 3
TROMBIDIOSIS (SCRUB-ITCH) INFESTATION BY CHIGGERS

Parkhurst (1937) gives a good account of immunological aspects on pp. 1027–8. The following are the major species of trombiculids causing scrub-itch. Some of these species are very local in distribution, e.g. *Schoengastia psorakari* to a part of east Malaya, but others are very widespread, e.g. *Eutrombicula wichmanni*, from Japan to Australasia and the Oriental region, is the counterpart of the almost-identical *E. splendens* of North America. The list is not complete: other species of local importance will surely be described.

Acomatacarus australiensis Wom. Australia.

Apolonia tigipioensis Torres and Braga, North Brazil (Pernambuco).

Blankaartia acuscutellaris (Gater). On one occasion, from a ricefield in Malaya.

Eutrombicula alfreddugesi (Ew.) (Synonyms are *Trombicula butantanesis, cinnabaris, irritans, lacertillae, rileyi, similis, tlalsahuatl* (?), *tropica, uruguayensis, vanommereni*). The chief pest-chigger of North America.

Eutrombicula batatas (Ew.) (Synonyms *T. braziliensis, flui, hominis, pastorae*) in South, Central, and southern North America; Jamaica, Puerto Rico.

Eutrombicula belkini Gould. U.S.A.

Eutrombicula goldii (Ouds.) (Synonyms *defecta, helleri*). U.S.A.

Eutrombicula samboni (Wom.) (Synonym *hirsti* Hirst). Ti-tree itch of South Australia.

Eutrombicula sarcina (Wom.) Very locally in Australia.

Eutrombicula splendens (Ew.) (Synonyms *masoni, scaber*). U.S.A. and Canada.

Eutrombicula wichmanni (Ouds.) (Synonym *T. buloloensis; minor* of Gunther, not Berlese) in Malaysia, Australasia (including Queensland) to Japan: very close to *E. splendens*.

Neoschoengastia nuniezi Hoffman, reported only from man in Mexico.

Neotrombicula autumnalis (Shaw). (Synonyms *T. holosericum, gymnopterorum, inopinatum, meridionale, pusillum.*) The major pest-chigger of Europe and Britain.

Neotrombicula desaleri (Methlagl). Austrian Alps.

Schoengastia psorakari Nadchatram and Gentry, 1964. Severe scrub-itch among local inhabitants on a sandy stream-bank, east coast of Malaya (*J. med. Ent.* **1**, 1–4, 1964).

Schoengastia schueffneri (Ouds.) (Synonym *S. pusilla*). Australasia and Sumatra. Military forces in Dobadura, Dutch New Guinea, suffered severely from scrub-itch, about 80% being due to *S. schueffneri* ('*pusilla*'), 15% *S. vandersandei* ('*blestowei*'), and 5% *E. wichmanni*. Recently reported (by James Gentry) as common around Kudat in Sabah (North Borneo).

Schoengastia vandersandei (Ouds.) (Synonym *yeomansi*) in Australasia. Vercammen-Grandjean is redescribing *S. blestowei* Gunther as a separate species although it has hitherto been considered a synonym of *vandersandei*. It remains to be confirmed whether one or both cause scrub-itch.

It is worth noting that several species have been recorded as causing severe reactions in animals. *E. wichmanni* (often accompanied by *Neoschoengastia gallinarum*) may be found in discrete ulcers on domestic fowls. Brennan and Yunker, 1964, report lesions on horses, dogs, and hares caused by *Euschoengastia latchmani*. Daniel, 1957, describes lesions on voles caused by *Cheladonta costulata* (Willman) (= *Eusch. ulcerofaciens* Daniel, synonym).

The following papers are particularly concerned with chiggers causing scrub-itch:

DANIEL, M. (1961). The bionomics and developmental cycle of some chiggers (*Acariformes, Trombiculidae*) in the Slovak Carpathians. *Czeskoslovenska Parasitologie* **8**, 31-118. [A very thorough study.]

DANIEL, M. and CERVA, J. L. (1964). Castosti vyskytu Trombikulosy ve Strednich Cechach [The frequency of occurrence of trombiculosis in Central Bohemia]. *Czeskoslovenska Parasitologie* **11**, 71-6.

FEIDER, Z. (1953). Citeva larve ale genului *Trombicula* (Acarieni) si descriverea unui caz de trombidiozu la sopirla *Lacerta agilis*. *Bull. sti.* **5**, 775-806.

KEPKA, O. (1962). Trombiculidae (Acari) aus der Türkei. *Z. Parasitenk.* **21**, 273-89.

KEPA, O. (1964). Die Trombiculinae (Acari, Trombiculidae) in Österreich. *Ibid.* **23**, 548-642. [Another very thorough study.]

KEPKA, O. (1964). Zur Taxonomie der Formen von *Neotrombicula* (*N.*) *autumnalis* (Shaw, 1790), (Acari, Trombiculidae). *Z. zool. Syst. Evolutionsforsch.* **2**, 123-73.

KEPKA, O. (1965). Merkblätter über angewandte Parasitenkunde und Schädlingsbekampfung. 12. Die Herbstmilbe (*Neotrombicula autumnalis*). *Angewandte Parasitologie* **6**, 1-13. [See Appendix 2.]

KEPKA, O. (1966). Trombiculidae (Acari) aus der Türkei. II. *Z. Parasitenk.* **27**, 43-63.

APPENDIX 3

MCCULLOCH, R. N. (1944). Notes on the habits and distribution of trombiculid mites in Queensland and New Guinea. *Med. J. Australia* ii, 543–5.

NADCHATRAM, M. and GENTRY, J. W. (1964). A new species of scrub-itch mite from Malaya, with notes on ecology (Acarina, Trombiculidae). *J. med. Ent.* **1**, 1–4.

OUDEMANS, A. C. (1912). Larven von Thrombidiidae und Erythraeidae. *Zool. Jahrbuch.* Suppl. **14**, 45–62. [See Appendix 2 for translations.]

PARKHURST, H. J. (1937). Trombidiosis (infestation with chiggers). *Arch. Dermatology & Syphilis* **35**, 1011–36. [Standard clinical account.]

PHILIP, C. B. (1964). Scrub typhus and scrub itch. Ch. XI, pp. 275–347 in *Preventive Medicine in World War II.* Vol. VII, *Communicable Diseases. Arthropodborne Diseases Other Than Malaria.* Office of Surgeon General, Department of the Army, Washington, D.C. [*Eutrombicula wichmanni* referred to as *T. minor*, habitat in Fig. 24, p. 308.]

POULSEN, P. A. (1957). *Undersogelser over Trombicula autumnalis Shaw og Trombidiosis i Danmark.* With English Summary. Denmark, Universitetsforlaget i Aarhus, 149 pp. [*Neotrombicula autumnalis* is the major European pest-chigger, the harvest-mite of Britain.]

TUXEN, S. L. (1950). The harvest mite, *Leptus autumnalis*, in Denmark, observations made in 1949. *Entomol. Medd.* **25**, 366–83.

VAN DER HAMMEN, L. (1956). Scrub typhus en scrub itch, in het bijzonder in Nieuw Guinea. *Zoologische Bijdragen* **2**, 1–55.

The following are the most useful references to the Trombiculidae generally and to particular species. References to major publications on chigger ecology in relation to epidemiology of scrub-typhus are in Appendix 5. The first 9 references are to systematic works including revisions of genera:

WHARTON, G. W. aided by H. S. FULLER (1952). *A Manual of the Chiggers. The biology, classification, distribution, and importance to man of the larvae of the family Trombiculidae (Acarina).* 185 pp. *Mem. Ent. Soc. Wash.*, Washington, D.C., U.S.A. (A basic work. The classification has since been greatly revised and many more species added.)

WOMERSLEY, H. (1952). The scrub typhus and scrub itch mites of the Asiatic-Pacific region. *Rec. S. Aust. Mus.* **10**, 1–435. (A large monograph with an obsolete classification, amended by H. WOMERSLEY and J. R. AUDY, 1957. The Trombiculidae (Acarina) of the Asiatic-Pacific region: A revised and annotated list of the species in Womersley, 1952, with descriptions of larvae and nymphs. *Stud. Inst. med. Res. Malaya* **28**, 231–94.)

AUDY, J. R. (1954). Notes on the taxonomy of trombiculid mites with description of a new subgenus. *Stud. Inst. med. Res. Malaya* (1953) **26**, 123–70.

APPENDIX 3

AUDY, J. R. (1957). A checklist of trombiculid mites of the Oriental and Australasian regions. *Parasitology* **47**, 217–94.

ZUMPT, F. (1961). The Arthropod Parasites of Vertebrates in Africa South of the Sahara (Ethiopian Region). Vol. 1 (Chelicerata). *S. Afr. Inst. med. Res.*, Johannesburg.

VERCAMMEN-GRANDJEAN, P. H. (1960). Introduction à une essai de classification rationelle des larves de Trombiculinae Ewing 1944 (Acarina-Trombiculidae). *Acarologia* **2**, 469–71.

VERCAMMEN-GRANDJEAN, P. H. (1963). Valuable taxonomic characters of Trombiculidae, including correlations between larvae and nymphs. *Advances in Acarology* **1**, 399–407. Cornell University.

VERCAMMEN-GRANDJEAN, P. H. (1966). Checklist of Acarina Trombiculidae, subfamily Trombiculinae. [Distributed to acarologists, with diagnoses, revised classification based on VERCAMMEN-GRANDJEAN, 1960, and with data on geographical distribution and hosts. Distributed in mimeographed form in 1965.]

VERCAMMEN-GRANDJEAN, P. H. and AUDY, J. R. (1965). Revision of the genus *Eutrombicula* Ewing 1938 (Acarina: Trombiculidae). *Acarologia* **7** (fasc. suppl.), 280–94.

WHARTON, G. W. (1946). Observations on *Ascoschöngastia indica* (Hirst 1915) (Acarinida: Trombiculidae). *Ecol. Monographs* **16**, 151–84. [*A. indica* is a nest-dwelling species in forests, now very widespread on house-rats from the Philippines to India and Australasia. It transmits *Rickettsia typhi* among rats.]

SASA, M. (1961). Biology of chiggers. *Annual Rev. Ent.* **6**, 221–4. [A general review of biology and distribution.]

SASA, M. (1956). [*Tsutsugamushi and Tsutsugamushi Disease*] (in Japanese). Igaku Shoin Ltd., Tokyo. [497 pp. An outstanding contribution.]

TAMIYA, T., Editor (1962.) *Recent Advances in Studies of Tsutsugamushi Disease in Japan*. Medical Culture Inc., Tokyo. [93 of 309 pages are devoted to chiggers and their control. Another outstanding work.]

KUMADA, N. (1959). Epidemiological studies on *Trombicula* (*Leptotrombidium*) *pallida* in Japan, with special reference to its geographical distribution, seasonal occurrence, and host-parasite relationships. *Bull. Tokyo med. & dent. Univ.* **6**, 267–91.

Appendix 4

JAPANESE AND OTHER NAMES FOR CHIGGERS AND SCRUB-TYPHUS

Japanese writing is exceptionally complex in using Chinese ideographs (*kanji*) combined with twice 109 characters in two syllabic phonetic alphabets, *hiragana* and *katakana*, the latter used for foreign words. There are no *ti, tu, si, zi* sounds in Japanese although they are written thus in the official *kunrei-siki* system of romanization established in 1937. In the alternative Hepburn system these same sounds are written as pronounced: *chi, tsu, shi, ji*. Thus *tutugamusi* and *Yamasita* are pronounced *tsutsugamushi* and *Yamashita*, and so written in the older Hepburn system. The more efficient official *kunrei-siki* makes sense only when one learns to read or write. There are also alternative pronunciations (*kun* and *on*) to many *kanji* ideographs, e.g., the ideograph for water is usually pronounced *mizu* but may be *sui* in some combinations as in *suison-netsu*, flood-fever; and the two ideographs for *tsutsuga-mushi* may be pronounced *yo-chu*.

Mushi is a fairly general term for insect or worm. The ideograph (see Chapter 2, page 36) is supposedly derived from a sketch of a worm. Its pronunciation as *mushi* has a Japanese origin but it may be pronounced *chu*, Chinese-fashion. *Dani* is more specific than *mushi*, never applicable to worms, but to tiny, smooth arthropods such as ticks (but not lice); hence *kedani*, the hairy-'tick' or chigger. Difficulties in translation arise because there is a tendency for general and very vague terms to acquire increasingly specific meanings without wholly losing their earlier vagueness. Many English equivalents now have fairly precise taxonomic meanings. Although the origin of bug is unknown, it seems to be akin to words such as *bogey* and *bugbear* stemming from a common root with the medieval English *bugge*, a scarecrow, and the modern Welsh *bwg*, a ghost. The idea conveyed is that of a nasty little creature, possibly harmful and not readily visible. But *bug* is now more precisely used for the beaked insects of the order Hemiptera. *Tick* comes to us through the German and has probably always been applied to any small, brown parasite which resembles the familiar acarines of the family Ixodidae. It has been applied equally to tick-like insects, such as hippoboscid flies and *Melophagus ovinus*, the sheep-ked. Possibly it has been applied to small creatures that attach and have to be pulled off. *Mite*, on the other hand, seems to have had the sense of something more tiny, an 'animated particle'. Since most arthropods of this size are indeed acarines, this word has probably always had a meaning close to its present taxonomic one. Herein lies a source of constant confusion, for we have now come to refer to the entire subclass Acarina by two common and mutually exclusive names, *viz. ticks*, referring to the Ixodoidea (hard and soft ticks), a relatively small and wholly parasitic group, and *mites*, referring to all the rest. But the Acarina comprise two major groups (see the classification in Appendix 1) that have very different origins: one including not only the highly specialized ticks, but also the many parasitoid mites; the other, much larger and much more varied, including such mites as the oribatids in the soil, the sarcoptiform mites including the cheese-mites and those causing scabies and mange, and the trombidiform mites including the chiggers of the family Trombiculidae. The parasitoid rat-mites are thus *very* much closer to ticks than to chiggers. As noted in Chapter 3, page 80, English usage of 'tick' and 'mite' has led to the grouping of rickettsialpox with mite-typhus (scrub-typhus) instead of with tick-typhus where it belongs. This confusion, due entirely to thoughtless use of words and a deficient popular vocabulary, may be found in some textbooks and in displays of teaching museums. The Russian word *kleshchi* (*-ei*, pl.) has the meaning of pincer or claw or something that nips. Although it applies equally to ticks and mites, it is usually translated as 'ticks' even though it refers to mites and this has led to some confusion in the literature; but the Russians would not have made the mistake described above. They now tend to qualify the word, e.g., *kleshchei krasnotelye* or red-*kleschei*, or simply *krasnotelki*, the red-bodied-ones, for trombiculids (although many trombiculids are white or at least pallid).

APPENDIX 4

The following are some basic words, followed in alphabetical order by the commoner terms used for scrub-typhus and for mites and ticks in the scrub-typhus region. Words suffixed to names to denote Prefecture, District are *-ken, -gun*.

Character	Romanized Term	Meaning
病	Byo	Disease, as in *tsutsugamushi-byo kedani-byo*.
虫	Mushi	Insect, 'bug', mite, small creature, worm. The ideograph for *mushi* is sometimes pronounced *chū*. *Jimushi* is a grub.
熱	Netsu	Heat, temperature, hence fever (*netsubyo* to be precise); also a craze or enthusiasm, as for baseball. Used in *kozui-netsu*, q.v.
蟲 or ダニ	Dani	Tick (or mite) originally vermin; not used for louse. Note the recurrence of the character for *mushi* in the ideograph. The same ideograph may be read Chinese-fashion as *shitsu*. See *Akadani-byo*.
赤ダニ病	Akadani-byo	Red-tick disease, scrub-typhus. *Dani* is often written in the phonetic Japanese alphabet (*Katakana*) thus: ダニ Most commonly used in Akita-ken.
赤虫	Akamushi	Red-bug; Niigata-ken.
赤虫 大明神	Akamushi-Daimyojin	Literally 'red-mite spirit(God)' (See *Daimyojin* and Figs. 8, 9.) A shrine set up to protect the people from the tsutsugamushi in a few parts of north-west Honshu, Japan, that are now

Character	Romanized Term	Meaning
		free of the disease (post-hoc evidence of efficacy). Also *kedani-daimyojin* (or *-myojin*). Tanaka (1962) states that the name may also be interpreted 'red-mites are enshrined' and the plaque shown in his fig. contains an elegant passage to this effect.
大明神	Daimyojin	God (or animus or spirit but not in the sense of a soul). *Dai*, great.
	Flood Fever	See Japanese River Fever.
	Flussfieber	River Fever. See Japanese River Fever.
	Gonone	Scrub-itch chiggers, New Guinea and adjacent islands.
	Japanese River Fever	Nagayo uses this term in 'The Tsutsugamushi Disease or Japanese River Fever' in English only. It is the equivalent of Baelz and Kawakami's *Flussfieber* (German equivalent of *Nihon kawa netsu*, q.v., used only exceptionally). These authors, together with Nagino, also referred to Flood Fever or *Überschwemmungsfieber* for which Japanese equivalents are *kozui-* and *suison-netsu*, q.v. It was evidently necessary in about 1878 to coin these terms because the role of the tsutsugamushi as a vector was temporarily denied and the fever ascribed to a miasma, in which case 'tsutsugamushi-disease' would be a misnomer.
毛蟲	Kedani	Hairy-tick, chigger; Akita-ken. See *dani*.

APPENDIX 4

Character	Romanized Term	Meaning
毛蟲 大明神	Kedani- Daimyojin (-Myojin)	See *Akamushi-Daimyojin*.
小蟲 or 小ダニ	Kodani	Small-tick or mite. See *dani*.
洪水熱	Kozui-netsu	Flood-fever (also *suison-netsu*, q.v., with the same meaning): a name coined by Kawakami and Baelz together with 'river fever', and rendered in German as *Überschwemmungsfieber*.
真蟲 or マダニ	Madani Mite-typhus Mokka	True-*dani* or tick. Proposed by Megaw. Scrub-itch chiggers, particularly *Eutrombicula wichmanni*, tropical Queensland and New Guinea.
日本河熱 or 日本川熱	Nihon kawa (kozui) netsu	Japan ('of' understood) river (flood) fever (Kakurai, 1915); Akita-ken. See *netsu*.
荻虫 or オギムシ	Ogi mushi	Tsutsugamushi, q.v., lit. grass-insect, *ogi* being *Miscanthus* grass, common in infected areas.
鬼棘 or オニトゲ	Onitoge	Term applied to chigger. Literally translated as 'the devil's thorn'. Commonly in Akita-ken. Usually written in the *katakana* phonetic alphabet reserved for foreign terms.
	Patau	Formosa: name given by aborigines to mite that causes disease, possibly scrub-typhus (Hatori, 1919).
	River Fever	See Japanese River Fever.
	Scrub Fever	Referred to by this name in the

Character	Romanized Term	Meaning
		Mount Molloy mining district of the North Queensland coastal range. CLARKE, *Appendix, Report of Commissioner of Public Health of Queensland for 1913*, summarized by A. BREINL, H. PRIESTLEY and J. W. FIELDING (1914). On the occurrence and pathology of endemic glandular fever, a specific fever occurring in the Mossman district of North Queensland. *Med. J. Austr.* **1**, 391–5.
	Scrub-Typhus	Name used in Malaya by Fletcher, Field and Lewthwaite. It came into general use during World War II.
沙虫	Shachu	Sand 'bug' (also *shashitsu*, q.v.) presumably either pest-chiggers or tsutsugamushi. Ri Tokuyu, a demoted prime minister of the To Dynasty in China, wrote a poem while travelling. In that poem entitled *Ryonan Dochu*, he says that 'the swallow avoids mud in fear of "shachu" ' according to Ogata, *Tokyo Iji Shinshi*, **70**, 52.
沙蝨	Shashitsu	Chiggers and probably other arthropods; lit. Sand-insect; Akita-ken. *Shitsu* is another (Chinese) way of reading the ideograph for *dani*, q.v.
七島熱	Shichito-netsu	Seven-island-fever. The islands of Izu, south of Tokyo. Fukuzumi confirmed that the fever was scrub-typhus. It is, however, a mild form transmitted

APPENDIX 4

Character	Romanized Term	Meaning
		in the autumn and winter by *L. scutellaris*.
島虫	Shimamushi	Island-mite, tsutsugamushi; Niigata-ken.
白虫	Shiromushi	White-mite, supposed by peasants to transmit scrub-typhus.
水損熱	Suison-netsu	Flood-fever. Used by Kawakami. Apparently he coined the word while working in Echigo, using it in the title of his report but using *kozui-netsu*, q.v., in the text. Quoted by Ogata (1953) *loc. cit.*, p. 61.
	Tlalzahuatl	Of Aztec origin, used in Mexico for pest-chiggers (*Eutrombicula batatas* and others). *Ahuatl* are the tiny and very irritating spines that occur on tubercles on the fruit of the prickly-pear cactus *Opuntia*. Touching spines embedded in the skin produces a sharp prickling sensation like that evoked by touching attached chiggers, hence the application of *tlalzahuatl* both to the chiggers and to the rash (trombidiosis or scrub-itch). Sambon (*Ann. trop. Med. Parasit.* **21**, 101, 1928) confuses the rash of trombidiosis (misspelt *tlalsahuatl*) with that of epidemic louse-borne typhus and mistakenly adduces the usage of this Aztec term as proving the transmission of typhus by mites, or at least to local beliefs to this effect.
恙ノ虫	Tsutsuga no mushi	Insect of danger—see Tsutsugamushi. The *no* is possessive.

APPENDIX 4

Character	Romanized Term	Meaning
		A common early dialect form, used, e.g., by Nagino.
恙虫	Tsutsugamushi	Insect (mite) of danger or illness (see also *Yochu-byo*). This is the most commonly used term but has been broadened to refer to trombiculids generally.
恙虫病	Tsutsugamushi-byo	Scrub-typhus: 'sickness of the noxious insect'.
	Tungau	Larval ticks and pest-chiggers, Malay.
	Überschwemmungsfieber	Flood Fever—see Japanese River Fever.
恙虫病	Yochu-byo	*Tsutsugamushi-byo* or dangerous bug-disease. See *byo* above. The ideographs are the same as for tsutsugamushi disease (simply another way of pronouncing the same characters). In 'The Tsutsugamushi Disease or Japanese River Fever' Nagayo gives *yochu-byo* as a synonym of *tsutsugamushi-byo* (p. 2134).

Appendix 5

MAJOR PUBLICATIONS ON SCRUB-TYPHUS AND ECOLOGY OF THE VECTORS

The following annotated references have been selected because they contain the major reviews of and advances in scrub-typhus research and, often, comprehensive bibliographies. The paper by Traub, Wisseman and Ahmad (1966) is an important contribution published since these Lectures were given.

In order to avoid confusion in these Lectures, the identification of the major vectors has been left unquestioned, although there is serious confusion in the identification, long taken for granted, of *L. akamushi*, *L. deliense*, and what has been identified as *L. deliense* in Asia. It is hoped that a neotype of *L. akamushi* will shortly be designated. *L. deliense* of Malaysia at least appears to be almost identical with *L. akamushi* of Japan, and *L. akamushi* of Malaysia may be yet another species. Dr P. H. Vercammen-Grandjean is presently sorting out these forms by comparison of material from many different places.

Refer to Appendix 3 for taxonomic and general accounts of trombiculids.

* The asterisk indicates broad investigations, reviews of the literature, and monographs.

*AUDY, J. R. (1949). Practical notes on scrub typhus in the field. *J. roy. Army Med. Corps* **93**, 273–88.

*AUDY, J. R. (1949). A summary topographical account of scrub typhus, 1908–1946. *Bull. Inst. med. Res., Malaya*, No. 1 (new series), 1–82. Summarizes the topographical and distributional aspects of wartime research in Imphal. Details of Indo-Burma region; 26 figs., maps and photographs. See also WAR OFFICE, 1947.

AUDY, J. R. (1956). The role of mite vectors in the natural history of scrub typhus. *Proc. 10th int. Congr. Entomol.* **3**, 639–49.

AUDY, J. R. (1959). The epidemiology of scrub typhus. *Proc. 6th Int. Congr. trop. Med. Malar., Lisbon* (1958) **5**, 625–30.

*AUDY, J. R. and HARRISON, J. L. (1951). A review of investigations on mite typhus in Burma and Malaya, 1945–50. *Trans. roy. Soc. trop. Med. Hyg.* **44**, 371–95. *cf.* Mackie *et al.*, 1946. Reviews work of British Scrub

Typhus Research Laboratory at Imphal, World War II—taking material from WAR OFFICE, 1947—and subsequent investigations in Malaya.

*BLAKE, F. G., MAXCY, K. F., SADUSK, J. F., KOHLS, G. M. and BELL, E. J. (1945). Studies on tsutsugamushi disease in New Guinea and adjacent islands: Epidemiology, clinical observations and etiology in the Dobodura area. *Amer. J. Hyg.* **41**, 243-73. Contains comprehensive review of literature with 152 references.

COOPER, W. C., LIEN, J. C., HSU, S. H. and CHEN, W. F. (1964). Scrub typhus in the Pescadores Islands: an epidemiologic and chemical study. *Amer. J. trop. Med. Hyg.* **13**, 833-8.

DAVIES, G. E., AUSTRIAN, R. C. and BELL, B. J. (1947). Observations on tsutsugamushi disease in Assam and Burma; the recovery of strains of *Rickettsia orientalis. Amer. J. Hyg.* **46**, 268-86. Includes some epidemiological observations.

DERRICK, E. H. (1961). The incidence and distribution of scrub typhus in North Queensland. *Ann. Med.* **10**, 256-67.

FARNER, D. S. and KATSAMPES, C. P. (1944). Tsutsugamushi disease. *Nav. med. Bull. Wash.* **43**, 800-36. Compact uncritical review with 201 references.

HARRISON, J. L. and AUDY, J. R. (1951). Hosts of the mite vectors of scrub typhus. I & II. *Ann. trop. Med. Parasitol.* **45**, 171-85, 186-94. Critical checklist of hosts and analysis.

*KAWAMURA, R. (1926). Studies on tsutsugamushi disease (Japanese flood fever). *Med. Bull. Univ. Cincinnati Coll.* **4** (Special Nos. 1, 2), 1-229. One of the classical studies.

*MACKIE, T. T., DAVIS, G. E., FULLER, H. S., KNAPP, J. A., STEINACKER, M. L., STAGER, K. E., TRAUB, R., JELLISON, W. L., MILLSPAUGH, D. D., AUSTRIAN, R. C., BELL, E. J., KOHLS, G. M., WEI, H. and GIRSHAM, J. A. V. (1946). Observations on tsutsugamushi disease (scrub typhus) in Assam and Burma. Preliminary report. *Amer. J. Hyg.* **43**, 195-218; and *Trans. roy. Soc. trop. Med. Hyg.* **40**, 15-46. Reviews work of U.S. Army Typhus Commission based on Myitkyina, North Burma, World War II. Relation of case incidence to the rainy seasons was masked by mass movements of troops in humid rain-forests, so that the number of cases seemed to be related only to movements of people.

MCCULLOCH, R. N. (1946). Studies in the control of scrub typhus. *Med. J. Aust.* May, 717-38. Classic account of use of DMP and DBP with troops.

MCCULLOCH, R. N. (1947). The adaptation of military scrub typhus mite control to civilian needs. *Med. J. Aust.* April, 449-52. Review plus field trial of gammexane; scrub itch mites in Australia.

*MEGAW, J. W. D. (1948). Scrub typhus investigations in South East Asia. *Trop. Dis. Bull.* **45**, 62-70. Abstract of War Office Report on work of the Imphal Laboratory.

MOHR, C. O. (1947). Notes on chiggers, rats and habitats on New Guinea and Luzon. *Ecology* **28**, 194-9.

*MOULTON, F. R., Editor (1948). *Rickettsial Diseases of Man*, Washington,

American Association for the Advancement of Science. Twenty-seven papers; 3 confined to scrub typhus; others on vectors, reservoirs, relationships, nomenclature, treatment and serology.

*OGATA, N. (1953). [Seventy-seven year history of research on tsutsugamushi disease in Japan] *Tokyo Iji Shinshi* [*Tokyo Med. J.*] **70**, 51–73. In Japanese, freely quoted in these Lectures and in Appendix 7.

PHILIP, C. B. (1947). Observations on tsutsugamushi disease (mite-born or scrub typhus) in dorthwest Honshu island, Japan, in the fall of 1945. *Amer. J. Hyg.* **46**, 45–59.

*PHILIP, C. B. (1948). Tsutsugamushi disease (scrub typhus) in World War II. *J. Parasitol.* **34**, 169–91. A valuable review of war experience, especially Pacific theatre; distribution map.

*PHILIP, C. B. (1964). Scrub typhus and scrub itch. Chapter XI, pp. 275–347 in *Preventive Medicine in World War II*. Vol. VII, *Communicable Diseases. Arthropodborne Diseases Other Than Malaria*. Washington, D.C. Office of the Surgeon General, Department of Army. Extensive and well-illustrated review.

PHILIP, C. B. and KOHLS, G. M. (1945). Studies on tsutsugamushi disease (scrub typhus, mite-borne typhus) in New Guinea and adjacent islands. *Amer. J. Hyg.* **42**, 195–203.

PHILIP, C. B., TRAUB, R. and SMADEL, J. E. (1949). Chloramphenicol (chloromycetin) in the chemoprophylaxis of scrub typhus (tsutsugamushi disease). I. Epidemiological observations on hyperendemic areas of scrub typhus in Malaya. *Amer. J. Hyg.* **50**, 63–74.

SASA, M. (1954). Comparative epidemiology of tsutsugamushi disease in Japan. *Jap. J. exper. Med.* **24**, 335–61.

*SASA, M. (1958). [Tsutsugamushi and tsutsugamushi disease], Igaku Shoin, Tokyo. Excellent survey in Japanese, especially of the taxonomy, biology, and distribution of Japanese trombiculids.

SAYERS, M. H. P. and HILL, I. G. W. (1948). The occurrence and identification of the typhus group of fevers in southeast Asia. *J. roy. Army Med. Corps.* **90**, 6–22. Important historical document.

SMADEL, J. E., WOODWARD, T. E., LEY, H. L., PHILIP, C. B., TRAUB, R., LEWTHWAITE, R. and SAVOOR, S. R. (1948). Chloromycetin in the treatment of scrub typhus, *Science* August, 160–61. Classic account of the first successful treatment of scrub typhus.

SOUTHCOTT, R. V. (1947). Observations on the epidemiology of tsutsugamushi disease in North Queensland. *Med. J. Aust.* **2**, 441–50. See also Derrick, 1961.

*TAMIYA, T., Editor (1962). *Recent Advances in Studies of Tsutsugamushi Disease in Japan*. Tokyo, Medical Culture Inc., 308 pp. Comprehensive survey of research after World War II. This account is complemented by the extensive reviews of Sasa (1954, 1958), *q.v.*

TANAKA, K. (1899). Über Aetiologie und Pathogenese der Kendanikrankheit. *Z. Bakt.* **26**, 432–9.

TRAUB, R. and FRICK, L. P. (1950). Chloramphenicol (Chloromycetin) in the chemoprophylaxis of scrub typhus (tsutsugamushi disease). V.

APPENDIX 5

Relations of number of vector mites in hyperendemic areas to infection rates in exposed volunteers. *Amer. J. Hyg.* **51**, 242–7.

TRAUB, R., FRICK, L. P. and DIERCKS, F. H. (1950). Observations on the occurrence of *Rickettsia tsutsugamushi* in rats and mites in the Malayan jungle. *Amer. J. Hyg.* **51**, 269–73.

*TRAUB, R., WISSEMAN, JR., C. L. and AHMAD, N. (1967). The occurrence of scrub typhus infection in unusual habitats in West Pakistan. *Trans. roy. Soc. trop. Med. Hyg.* **61**, 23–57. Published while the present Lectures were in press. Describes what may be relict islands of infection, transmitted by species of *Leptotrombidium*, in desert and remote high mountainous areas.

*WAR OFFICE, Army Medical Directorate, Great Britain (J. R. AUDY *et al.*) (1947). *Scrub Typhus Investigations in South East Asia. A Report on Investigations by the G.H.Q. (India) Field Typhus Research Team, and the Medical Research Council Field Typhus Team, based on the Scrub Typhus Research Laboratory, South East Asia Command, Imphal.* Restricted distribution; abstracted in *Trop. Dis. Bull.* **45**, 62–70, 1948. Includes reviews, 22 papers, 135 illustrations, on investigations in Manipur and Burma. Data on relation to seasonal rainfall or perennial ground-water; life-cycle of *L. deliense* in nature; feeding-times of chiggers; 'ecological labels'; relationships to vegetation patterns and 'fringe-habitats'. Collaborators were: J. R. Audy, J. D. Bower, H. C. Browning, A. A. Bullock, K. L. Cockings, W. Ford, H. T. M. Gordon, S. L. Kalra, M. T. Parker, C. D. Radford, M. L. Roonwal, H. M. Thomas, Reviews by M. P. H. Sayers and I. G. W. Hill (*q.v.*), R. Lewthwaite, and J. R. Audy.

Appendix 6

A NOTE ON MIASMA AND MOSQUITOES AS CAUSES OF DISEASE

Baelz was faced with some difficulties in proposing that miasma causes scrub-typhus and his explanations are perhaps the only feature that distinguishes his paper from those of Nagino and Kawakami. This may be construed as evidence that he, more than the others, favoured the miasma hypothesis. Why did farmers not get *malaria* from the miasmas? Why did the river fever not occur in *other* rivers with grassy banks? He invokes 'another specific cause'. Why do boatmen on the same rivers not get infected? Because (he says) it is the cutting of hemp that exposes one to miasmas from the soil. He notes that all day long farmers stand in a bent-down position and expose their bodies to the miasmas rising from the ploughed-up soil. Why do some patients insist that they have never been exposed on the alluvial *yudokuchi*? Because, Baelz claims, the poison is transportable with the hemp and the earth adhering to its roots—and here he could be right by accident, because we know that vector mites can be carried into buildings with straw and fodder. His arguments should be studied in the original, in detail. They show how doggedly a scientist will defend his hypothesis. There are many examples of such activities today and I would presumably defend my belief in the soil-to-trombiculids origin of the typhus group of rickettsiae similarly.

Baelz does admit that we may be dealing with a *miasma vivum* or *inficiens vivum*. Sternberg in 1884 started his book on malaria thus:

> The malarial diseases are included by Liebermeister in his group of 'Infectious Diseases', which, according to his definition, originate through the infection of the system with certain peculiar poisonous matters, which are mainly distinguished from the ordinary poisons by the fact that they can reproduce themselves under favoring

conditions to an endless degree. They also fall within the sub-group of 'miasmatic infectious diseases', in which the poison *develops itself* externally; its reception into a higher organism not being necessary to its reproduction.

Now to admit that the malarial poison reproduces itself external to the body would be to concede, at the outset, that it is a living germ. This we are not prepared to do, and consequently place at the head of this chapter, 'Mode of Infection or Intoxication(?)'.

There can be no question as to the manner in which malarial poisoning usually occurs. The term 'malaria' indicates the usual belief that it is due to contamination of the atmosphere—bad air; and a multitude of facts support this belief. A brief exposure in such a contaminated atmosphere, in intensely malarious regions, is often sufficient to produce poisoning of the most serious character; and it is well established that in such regions the atmosphere is more highly charged with malaria at night than during the daytime; that the poison is more intense near the ground than at a little elevation above the surface from which it is evolved; that it may be carried by the winds from the locality where it is produced, etc.

I will not labour the important lessons to be learned from such accounts. There is a sort of arrogance about many scientists that should be corrected. Perhaps science lends itself to arrogance among those who are not sufficiently close to nature to learn humility.

Yet many people before Baelz and the Japanese farmers had believed that insects could transmit diseases. Rupert W. Boyce (*Mosquito or Man?* 3rd ed. London, John Murphy, 1910, pp. 22–3) says:

There appears strong evidence that the danger of flies and mosquitoes was known in very early times. Thus Sir Henry Blake, in speaking at a banquet in connection with the Liverpool School of Tropical Medicine in 1908, mentioned how, when he was Governor of Ceylon, he had been shown a medical work written fourteen hundred years ago, in which the mosquito was stated to be a carrier of disease, and malaria was described as being transmitted by flies or mosquitoes. It will be remembered that Herodotus spoke of winged serpents. Beauperthery argues that this term is very applicable to mosquitoes, whose poisonous bite he compares in its effects on the human body to that of the serpent's bite. The use of mechanical protection against mosquitoes also appears to be a very ancient practice, the means adopted consisting of either smearing the exposed parts with pungent fats and oils, or more commonly by the use of netting; this is seen in the use of our common word 'canopy' (κώνωψ gnat).

Appendix 7

NOTES AND TRANSLATED EXCERPTS FROM JAPANESE PUBLICATIONS

Some of the early Japanese literature on tsutsugamushi disease is very inaccessible, even to students in Japan, as Dr Norio Ogata testifies in the paper referred to below. The following are a few excerpts, from Japanese publications, to supplement the passages quoted in the text of these lectures.

OGATA, Norio (1953) [Seventy-seven year history of research on tsutsugamushi disease in Japan] *Tokyo Iji Shinshi* [*Tokyo med. J.*] **70** (10), 51–73 (see acknowledgements on page 29).

There are a few inconsistencies and confusing passages in this article. Some of these are due to difficulty in separating the author's own text from that of verbatim quotations of other authors (mostly Seiya Kawakami).

Part I: *Introduction*
See quotations on pages 39, 46, 49.

Part II: *Derivation of 'tsutsuga'*
'An envoy [from China] was sent to Japan in the Zui dynasty [Sui in Chinese; sixth century] and said, "The Emperor of the land of the Rising Sun sends a letter to the Emperor of the Land of the Setting Sun; have you been tsutsuga-nake? [i.e. without danger] ...

'The disease "shashitsu" is described in old Chinese books: in the *Hobokushi* [Pao p'o tsu 抱朴子] by Kakko [Ko, Hung 葛洪] in the *Hitsukoho* [*She hou fang* 射後方] by Tokokei [T'ao, Hung-ching 陶弘景], and in the *Byogen Koron* [Ping yüan hou lun 病源候論] by Sogempo [Ch'ao, Yüan-fang 巢元方] ...'

Ogata also quotes Li Shi-Chên's description of scrub-itch found in *Honsomo Mokusubu*, vol. 42 (In Japanese, *Honzo Komoku* by Rijichin). See asol pp. 37–38.

Part III: *Up-to-date information relating to the tsutsugamushi disease and the tsutsuga mite*

This is a clinical and diagnostic account which is no longer worth quoting. Ogata refers, in a special section, to cases of laboratory infections and pays homage to those who died from these. Shoichi Kitagawa was the first victim in October 1927, in Chiba City. Infection was caused by a prick of the finger; fever ensued some three weeks later; and death, on the eleventh day of illness. In 1952, Akiyoshi Kawamura, also in Ogata's laboratory, contracted the disease. 'In this twenty-five year period, cases of infection in various laboratories of Japan probably amounted to 50–60, among them fewer than ten died' (a remarkably low mortality rate for such infections). Ogata names Sugata, Kashima, Professor Nishibe, and the following students who were still alive after laboratory infections: Shiro Kasahara, Tokichi Musorida, Sadakichi Fukuzumi, Mononori Nakajima.

Ogata then discusses the controversial issue of priority in the discovery of the rickettsia. 'This is like gambling: one wins and all the others lose' (see page 54). The correct name for the rickettsia and its authorship are discussed. This, however, depends on purely technical aspects involving priorities and the issue is therefore, for good reasons, associated with strong feelings among Japanese scientists, who favour *R. orientalis* rather than the *R. tsutsugamushi* regarded as official in Bergey's Manual. (Is this because accepting the latter name would immediately rouse fierce controversy as to the correct author of the specific name?) There follow notes on the tsutsugamushi, the field-rodents on which they feed, and the 'areas of occurrence' in parts of the three prefectures, Niigata, Yamagata, and Akita. 'These areas are along rivers which flood every year. For this reason, the disease was sometimes named "flood-fever".' However, as far as I have been able to discover, this name, like that of 'river fever', was coined at the time when doubt was cast on the idea of transmission by the tsutsugamushi (see Chapter 2).

Jizo temples and rituals are then discussed in some detail (see pp. 34, 39), followed by a 'brief history of tsutsugamushi and research thereon'. The latter mostly concerns the causative agent.

APPENDIX 7

Part IV: *Individuals who have devoted themselves to the study of the tsutsugamushi disease and the tsutsugamushi in Japan*

Ogata quotes from Nagino's 1878 report, and he gives historical information about Nagino, Kawakami, Baelz, and also the following workers on the tsutsugamushi (here alphabetical): Norihiko Asakawa, Shigeo Hatori, Kiyoshi Hayakawa, Naosuke Hayashi, Arao Imamura, Taichi Kitajima, Shibasaburo Kitasato, Masao Kume, Tokushiro Mitamura, Keinosuke Miyairi, Mikinosuke Miyajima, Yoneji Miyakawa, Matao Nagayo, Masujiro Nishibe, Masanori Ogata, Norio Ogata, Masatada Okumura, Manabu Sasa, Kiyoshi Sato, Tomoyoshi Sawada, Takeo Tamiya, Keisuke Tanaka, Masanori Taramura, and Jiro Ukai.

Part V: *Progress and results of studies on tsutsugamushi disease in Japan*

Here Ogata gives his musings on rivalry and research (see quotation on pages 54–55).

Part VI: *Conclusion*

This section includes a summary and a repeated homage to those who devoted or even sacrificed their lives to research on the tsutsugamushi disease.

KAKURAI, Tokibumi (1915) [*Tsutsugamushi disease in Yamagata Prefecture*] Tokyo Printing Co. for the Police Department of the Yamagata Prefecture (58 pages).

Dr Nobuo Kumada (see page 35) is to be credited with discovering this hitherto-overlooked pamphlet written for the instruction of the public as well as of doctors by Kakurai, a health officer or sanitarian of the Yamagata Prefecture. The text, frequently repetitive, is a curious mixture of technical data and extremely simple instructions, written in old-fashioned Japanese. There is a brief introduction by Professor Miyajima of the famous Kitasato Research Institute.

Page 1: 'In the twelfth year of the Meiji era [1879] Baelz made the disease known under the name of Japanese River or Flood Fever [*Nihon kawa netsu; Nihon kozui netsu*].' This tends to confirm the opinion given on page 52. See also pages 48, 157.

APPENDIX 7

Page 2: Kakurai refers to two endemic areas in Karenko, east Formosa [discussed by Hatori, 1919], 'namely Muh Gua near the bank of a river and Feng Lin in a forested area ...' (see page 53). The local forms of scrub typhus were named by adding *Reh* (fever) to these places.

Pages 3–4: 'In Nishimuraya district ... this disease appeared in the forty-first year of Meiji [1908] ... In the second year of Taisho [1913] a man working in the recently cultivated area became ill ... This was the first case that finally led to the discovery of the tsutsugamushi disease ... by Drs Kitajima and Tanaka ...' This presumably means that the case led to the definite diagnosis of tsutsugamushi disease in that particular locality. Sixteen cases with nine deaths for that year are tabulated.

Page 5: Tabulation of cases and deaths from 1908 to 1914.

ISHII, Hitoshi (December, 1965) [Close-of-the-year ceremonies rich in local colour: The Shōrei festival of Mt Haguro] *Tabi* [*Travel*] **39** (12), 116–17.

There are references in Chapter 2 to ceremonies related to the tsutsugamushi, including one at Mt Haguro (Fig. 5). In the month after these Heath Clark lectures were given, an interesting account appeared in a travel magazine, *Tabi*, illustrated by a photograph of a group standing around the burning effigy of the tsutsugamushi at the Haguroyama Shoreisai (Mt Haguro Shōrei Festival) in Yamagata Prefecture. Mt Haguro, 419 metres, is one of a chain called the Three Mountains of Dewa, where an ascetic cult of mountain priests has been developed since the Heian Era [eighth to twelfth century]. The following is a précis of the rambling and repetitive article.

The most typical of all festivals celebrated in the Mt Haguro Dewa Temple is the Shōrei Festival that continues from New Year's Eve till dawn. The leaders of this festival are called *Matsuhijiri* (lit. 'sacred pine'), of whom there are two: Ijō and Sendō. These two undergo purification rites for a hundred days prior to New Year's Eve, when they are finally ready to compete against each other for the favours of Buddha. Near Mt Haguro are two villages, 'Upper Village' and 'Lower Village', separated from each other by a creek. Following tradition, the Upper Villagers support Ijō and the Lower Villagers Sendō.

On 31 December, the Matsuhijiri lead the youths of these villages to the temple on Mt Haguro, where they participate in a series of rites whose outcome will predict the year's success in harvesting and fishing. They first cut about 50 cm off the rope binding the *tsutsugamushi*, a large *taimatsu* [lit. 'burning pine', a torch made of grass and rope], which they shred and scatter on the ground. As these shreds are believed to prevent fire, pilgrims to the temple vie with each other in collecting them. Then in *Tsuna-sabaki*, a ritualized tug-of-war, the two competing groups of young men test the strength of the ropes that will be used to haul the two *tsutsugamushi* into shallow holes that have been dug by eight youths in competition. When a conch is blown, the main event takes place: the men, led by Ijō and Sendō, run towards the ropes, tie them around the large *tsutsugamushi*, which they drag into the holes and burn. The *tsutsugamushi* that burns more quickly and brightly decides the winning side and augurs well for the village concerned.

Appendix 8

TYPES OF OUTBREAKS OF SCRUB-TYPHUS

The following are examples of types of case-incidence of scrub-typhus that are related to human behaviour.

(1) *The 'explosive' outbreak on first encountering the yudokuchi*: All outbreaks of scrub-typhus are due to exposure of a body of men to some infected focus or *yudokuchi*. Sometimes these outbreaks reach epidemic proportions and they have, indeed, been called epidemics. An epidemic, however, is the spread of an infection among and between men and it always builds up and subsides in characteristic ways decided by the means of transmission of the organism, the susceptibility and density of the population, and the incubation period. In contrast, the size of an outbreak of scrub-typhus is decided entirely by the number of people who happen to expose themselves to the infected focus at one time. The incubation period of scrub-typhus is about 10-12 days so that nothing is noted for the first week or so after exposure. The body of men concerned may often be in a different place at the time the outbreak strikes them and this kind of infection *en passage*, characteristic of estate labourers in peacetime, is very common amongst military units. In Burma, one unit passed through an infected focus on its way to attack Kalewa and suffered over 80 cases of scrub-typhus coinciding with their arrival at their military objective. One incident at the tin mining camp at Mawchi, east of Toungoo on the Shan plateau, resembled the circumstances encountered by estate labourers clearing undergrowth. For 10 days, a single company cleared the overgrown compound around the mining-camp living-quarters, which were badly neglected and overrun by rats. They then moved on down the hillside clearing rank grass from under a power line for a further 8 days. From the 13th to the 17th day after starting, 7 cases of scrub-typhus developed, all obviously traceable to the first area around the living quarters. Other companies engaged in clearing the hospital area and the roadsides were unaffected,

although roadside grass is known to be frequently infected. A unit 700 strong started in mid-June, 1945, to clear an overgrown area in Prome, which for three years had been a derelict bombed-out township, and it had 11 cases of scrub-typhus. The interesting point about this outbreak is that 2 weeks beforehand the same unit had cleared a similar area only a few hundred yards away without a single case of scrub-typhus. This illustrates clearly the patchiness of the infection.

(2) *Rapid decline after occupation:* Clearing and 'civilizing' infected areas led to sharp and characteristically declining outbreaks. A unit 360 strong cleared an open grassy site on the banks of the Irrawaddy, opposite Kalewa, in September 1944 and suffered a weekly incidence of 6, 2, 2, 2, 1. A unit only 34 strong, which cleared and occupied an infected compound in Fort Dufferin at Mandalay, reported 6 cases all within 5 days. This bungalow and its compound became known as 'Typhus House' after 3 successive outbreaks.

(3) *Sporadic cases following occupation:* Scrub-typhus casualties may continue to appear wherever camps are improperly cleared or whenever men frequent the infected perimeter. A unit 300 strong in Prome had 20 cases spread over no less than 16 weeks, apparently due to exposure in adjacent neglected native gardens and the old market area. Incidentally, this is the sort of incidence to be expected of flea-borne typhus, especially if anti-rat measures are adopted in the buildings, and the fleas are deprived of their favourite hosts.

(4) *Delayed incidence:* Philip (1964: 289) has described 3 outbreaks in which there was a peculiar delay of some weeks after the units were installed. (a) *Milne Bay*, New Guinea: 9 cases between the third and fourth months. (b) *Goodenough Island:* No cases on clearing 2 hospital sites, which moved in during September–October 1943; 48 cases in patients and staff of 2 hospitals in November and December, 24 being in the latter half of December in one of the hospitals. (c) Near *Finschhafen*, New Guinea: 17 cases in 4 batteries 88–112 days after occupying a well-cleared and already-'civilized' site. (The last two outbreaks had exceptionally high fatality rates, respectively 27·5 per cent and 35·3 per cent.) In addition, Audy (1949: 20) described a similar episode first reported by D.

Tomlin: (d) *Myitnge*, Mandalay district, Central Burma: no cases in a unit 350 strong on clearing and occupation of a riverside camp, 1 October 1945; 7 cases between 25 October and 7 November.

The most likely explanation of such episodes is the discovery and use of a *yudokuchi* that had hitherto escaped attention. This was apparently the case in (c) and (d). The Finschhafen outbreak was traced to the clearing for and attendance at a new grassy amphitheatre in a nearby ravine (Philip's Fig. 16, p. 90). This theatre had ironically been named after T. J. Ayres, the first scrub-typhus victim. The Myitnge outbreak was traced to foraging and defecating in an adjacent bombed-out village that had become completely overgrown.

Other explanations that may be invoked include settlement during an off-season (dry season in the case of *L. deliense* and *L. akamushi*) and an outbreak, for example, after the rains start; also an infected patch of waste land may be included within an occupied area and remain undetected until it is later used or cleared.

(5) *Repeated reoccupation:* Evidence from many scources suggested an increase in incidence of scrub-typhus where camp-sites were repeatedly abandoned and reoccupied (Audy 1947). A camp is always attractive to rats but abandonment leads to the growth of grassy cover within the camp and this must be cleared on reoccupation, a risky procedure.

AUDY, J. R. (1947). An ecological study of scrub typhus, pp. 15–71 in Vol. 1 of report, *Scrub Typhus Investigations in South East Asia*, London, War Office (see *Trop. Dis. Bull.* **45**, 62–9, p. 65).

AUDY, J. R. (1949). A summary topographical account of scrub typhus. *Bull. Inst. med. Res. Malaya* (new series) **1**, 1–82.

PHILIP, C. B. (1964). Scrub typhus and scrub itch. Chap. XI, pp. 275–347 in *Preventive Medicine in World War II*. Vol. VII, Washington, D.C., Office of Surgeon General, Department of the Army.

Appendix 9

MASS FLOWERING OF BAMBOOS AND SOME CONSEQUENCES

In many places, bamboos are vigorous weeds, replacing the forest as it is cut down by shifting cultivation until hundreds of square miles are covered by impenetrable bamboo. These gregarious bamboos flower at long intervals of 50–100 or more years. The flowering is either simultaneous over an entire bamboo tract, or it spreads in a rapidly expanding wave, after which fruit is borne and the whole tract dies down while new young growth springs through into the sun. Some bamboos, especially the *Melocanna* in hills of the southern Indo-Burma border, produce large quantities of nutritious seed eaten by almost every available animal, especially rodents but including elephants. It is usual for a mass-flowering to be followed by a plague of rats swarming over the neighbouring countryside, destroying crops and invading homesteads. Famine follows, and villages are often abandoned after such an event.

For example, in 1958 a deputation of Lushai on the Indo-Burma border near Imphal warned the local administration that there would be famine in 1959–1960 because the bamboo was due to flower. The last time, in 1908, was followed the next year by a famine during which many Lushai died in their lonely hilltop villages. The administration shrugged its shoulders over the warning in 1958. How could the Lushai know such a thing? Why should they listen to an old wives' tale? (It is the habit of bureaucracies to ignore dire warnings.) But, awakening too late, when Lushai were already dying of starvation, the Government committed itself to over £1·8 million of expense in 1960 (according to the news report) in order to try to feed the Lushai by air. Every attempt by the Government to get the Lushai to leave their villages so as to be fed at selected valleys by airdrops failed; for the Lushai have a tradition of living safely in their hilltop villages. They felt, and would doubtless again feel, that it is the responsibility of the Government to feed them, especially after fair warning that famine was due.

Indeed, the fact that their warnings had been treated lightly added new resentment and made them less co-operative.

TROUPE, R. S. (1921). The silviculture of Indian trees, **3**, 977. Oxford University Press.
ZINKIN, T. (1960). Death and bamboo flower. Starvation in Lushai Hills. *The [Manchester] Guardian.* 18 February 1960. (Dispatch from Bombay, 16 February.)

Appendix 10

PREVENTION AND TREATMENT

Great success was achieved during the war in personal protection by impregnating clothing with mite-poisons such as dimethyl or dibutylphthalate (DMP, DBP) or benzyl benzoate —to which we can now add many others. An experimental vaccine was prepared from the lungs of infected cotton-rats as an emergency measure but in tests made after the war the course of the disease in vaccinated persons who later contracted the disease showed that the vaccine offered but little protection. There are several distinct strains of the rickettsia and it would be necessary to make a polyvalent vaccine. Workers in Japan now believe that they have advanced far enough to do this for the protection of people at least in Japan.

The most dramatic advance in typhus research generally was the post-war discovery of potent antibiotics that would bring a seriously ill patient's temperature down in a matter of hours. The first on the scene was chloromycetin, since synthesized and now officially known as chloramphenicol. The story of chloromycetin is a splendid tribute to team work. Prof. Paul R. Burkholder at Yale University selected a few moulds that showed some antibiotic activity from among some 20000 moulds grown from 6000 soil samples from every corner of the world.

Prof. M. A. Joslyn under the direction of J. Ehrlich of Parke Davis & Co. found that an antibiotic from a *Streptomyces* from Venezuela was active against rickettsiae and a number of other organisms. Dr Q. R. Bartz and his colleagues isolated the antibiotic in its pure crystalline form. The Institute for Medical Research in Kuala Lumpur in 1948 was an obvious place for trials of the new drug, and an American team duly arrived carrying the entire world supply of chloromycetin (then about 2 lb.). This team was led with inspiration and great vigour by Dr Joseph E. Smadel, at that time Chief of the Division of Virus and Rickettsial Diseases of the Walter Reed Army Medical Department Research and Graduate School

28. AMONG THOSE PRESENT ...

		WOODWARD	COCKINGS		PHILIP	
		(US)			(US)	
TRAUB	LEY			SAVOOR		LEWTHWAITE
(US)	(US)					
		AUDY	SMADEL		HARRISON	
			(US)			

Fig. 13. Laying the bogey of scrub-typhus to rest in Malaya in 1949 at the successful conclusion of trials of chloromycetin (chloramphenicol), the first drug ever to cure typhus infections. By artistic license, Audy, Cockings, Harrison, and Savoor are given undue prominence in this effort (but not the celebration!), which was that of the late Joseph E. Smadel's team. Cartoon by Austin Dorall (then Institute for Medical Research). Historical review by various authors, page 83: *The Institute for Medical Research, 1900–1950.* (*Stud. Inst. med. Res., Malaya*, No. 25 (Jubilee volume): 389 pages. Kuala Lumpur, Government Press, 1951.)

(now WRAIR) at Washington, D.C., Within a few days the efficacy of chloromycetin in the treatment of scrub-typhus was demonstrated beyond doubt. Shortly, Dr H. M. Crooks, Jr.

and Dr M. C. Rebstock determined the chemical structure of chloromycetin and, in collaboration with Dr John Controulis, all of Parke, Davis and Co., succeeded in synthesizing it. Hard on the heels of chloromycetin (chloramphenicol) came the equally efficacious tetracyclines.

Among the first people treated in Malaya with chloromycetin was a patient suffering from typhoid (enteric fever). He, and others with typhoid after him, responded equally well to the new drug. The American team returned for further studies and finally became established in newly-built quarters as the U.S. Army Medical Research Unit which is still at work in the 'I.M.R.' at Kuala Lumpur. Mr A. Dorall, administrative assistant to Dr Lewthwaite, Director of the 'I.M.R.' at that time, drew the cartoon reproduced in Fig. 13, representing the laying to rest of the bogey of scrub-typhus. Although I am shown in this cartoon with my colleagues John Harrison and Kenneth Cockings, we only collaborated as guinea-pigs and in small ways in field-work; the credit for laying the bogey belongs to Dr Smadel and his team, and to those who found and isolated the antibiotic.

INDEX

(Place names usually refer to outbreaks of scrub-typhus)

abandoned villages, 108
Abyssinia, infected ticks, 40
Acarina, classification, 126
—, typhus vectors, 79
acarine dermatoses, 147
acariosis, 147
achievements, recognition, 56
actinochitin, 126
adaptation, parasites, 13
Addu Atoll, 98
Admiralties, Bat I., 99
African view, prevention of relapsing fever, 41
air photography, uses, 106, 107
akamushi, i (Fig. 1), 28, 38, 45, 53, 155
—, life-cycle, 3 (Fig. 2)
—, without infection, 48
—, shrines to, 32, 33 (Fig. 8), 34 (Fig. 9)
Akamushi-Daimyojin, 34 (Fig. 9), 155
Akita Prefecture, 28, 34, 39, 52
Allodermanyssus sanguineus, 80
amaeru, Japanese psychology, 59
America, pest-chiggers in, 78
Anigstein, Ludwig, 74, 106
animal, laboratory, search for, 74
animal weeds, 25, 82, 91, 93
anthropocentric attitudes, 68
Apolonia tigipioensis, 149
Arakan Yomas, 85
Arashi, shrine at, 35
Arato, shrine at, 32
arthropods, pestiferous, 9, 128, 143
—, transmission of diseases, 40
Asanuma, Kiyoshi, viii, 33, 34
Ascoschoengastia indica, 14
Asia, distribution chiggers in, 78
Ash, G. W., 101
Audy, J. R., 6 (Fig. 4), 76, 179 (Fig. 13)
Audy's Circus, 85
Australasia, scrub-typhus, 28

Australia, 28, 102
Austria, scrub-itch, 150
autecology, 24
autumnal erythema, 10
Ayres, T. J., 175

Bacon, Roger, 21
Baelz, Erwin, 43, 44, 45, 51, 56
—, biography, 45
—, diary, 49, 56
Baker, Henry, 1, 142
balance of nature, 91
bamboo, mass flowering, 176
—, tree-mouse, 14
Bartz, Q. R., 178
Bat I., 99
bats, and *Trombicula*, 39
bee-stings, 16
behaviour of hosts, and chiggers, 12, 13, 16
— of ungulates and man, affected by arthropods, 145
Benedict, Ruth, 59
benzyl benzoate, 2, 10
biocenology, 25
bionomics, 24, 25
birds, roles of, 15, 116
— and scrub-itch, 131
Bishenpur track, 86
Bitter Root Valley, 65
Blankaartia acuscutellaris, 14, 149
blood-sucking, by chiggers, 47
Borrelia, 114
Boston, rickettsialpox, 114
bottle, wrongly labelled, 71
Brill's (Brill-Zinsser) disease, 66
Britain, chiggers in, 15
British Army Laboratory, Imphal, 85, 97
British Typhus Commission, 97
Browning, H. C., 101
Brumpt, E., 39
'bug', origin, 153

INDEX

Bullock, A. A., 101
Burma, main routes into, 85
—, refugees from, 86
Burmese eruptive fever, 110
burning of mite effigies, 31
Burkholder, Paul R., 178
Burton, Richard, 40
bush-turkey, 10, 17, 130

Calcutta, 87
Capillaria hepatica, 122
Carson, Rachel, 25, 57
Cheladonta costulata, 150
ceremony, Hagura, 30
—, ticket-throwing, 35
Ceylon, 87, 88
Chagos Archipelago, 98, 116
Changi camp, Singapore, 119
changing patterns of disease, 113
Charts, life-cycle, 3 (Fig. 2), 4 (Fig. 3), 6 (Fig. 4)
chiggers, 8, 9
—, and behaviour hosts, 12, 13, 16
—, Addu Atoll, 13, 14
—, derivation, 8
—, feeding of, 90, 117, 132
—, Japanese names for, 153
—, protection from, 100
—, re-attachment of, 117
—, and scrub itch, 15, 145, 149
chigoe flea, 8
Chinese herbal, 37
Chindwin, 86
Chiropodomys, 14
Circus, Audy's or Imphal, 85, 97
chloramphenicol, 178
chloromycetin, 178
classical accounts, biasing, 96, 110
classification of acarines, 126
climax vegetation, 92
clothing, treatment of, 100
coastal fever, 75
Cockings, K. L., 100, 179 (Fig.13), 180
community ecology, 25
competition, scientific, 55
confusion, names of vectors, 104, 161

Controulis, John, 180
controversy about vector, 102
—about vegetation, 106
Cook, C. E., 102
copra itch, 148
Coxiella burneti, 77, 78
crabs on Jarak, 120
criticism, scientific, 55
Crooks, H. M., 179
cultivation, shifting, 88, 90, Plate I (Fig. 11), Plate II (Fig. 12)
curiosity, 19
Czechoslovakia, scrub-itch, 10, 15, 150

Daintree River, 66
dani, 153
Daniel, M., 150
data-processing in education, 21
Davis, Gordon, 99
DBP, 100, 178
declining incidence, 174
deer, influenced by flies, 145
delayed incidence, 174
deliensis, *Trombicula*, 69 *and see Leptotrombidium*
Denmark, 15, 18, 151
Dermanyssidae, 148
dermatoses, acarine, 147
dibutylphthalate, 100, 178
Diego Garcia, 98, 116
Dimapur (Manipur Rd.), 86
dimethylphthalate, 100, 178
diseases, 'new', 115, 116
—, arthropod transmission, 40
dispersal of vectors, 96
distribution, chiggers, 10, 11
—, effects on animal, 9, 143
—, parasites, 12
—, scrub-typhus, 42 (Fig. 10), 81, Plate II (Fig. 12)
DMP, 100, 178
Doi, Takeo, 58
domestic waste land, 108
Dorall, Austin, 179 (Fig. 13), 180
drinking-straws of chiggers, 16

INDEX

East Africa, names for vectors, 40
Easter, Avis, viii
ecological labels, 12, 13
—'slant', 26
ecology and bionomics, 24
—, nature and scope, 22, 24, 25
ecosystem, 25, 91
edge-effects, 107
education, 20, 21, 23
effigies of tsutsuga, 30 (Fig. 5), 31 (Fig. 6), 32 (Fig. 7) 172,
Ehrlich, J., 178
Elton, Charles, 93
Embilipitiya, 87, 88, 97
encephalitis, equine, 116
endemic areas, Japan, 42 (Fig. 10), 43
— region, 12
— typhus, 64, 65
endemicity, increasing, 101
Ensoll, Ben, 120
enteric fever, 63
environment and ecology, 24, 26
epidemic typhus, 63, 64, 65, 74, 77, 78 (Table I)
equine encephalitis, 116
epidemiological types of terrain, 108
Erythraeidae, 148
Ethiopia, infected ticks, 40
Europe, scrub-itch, 11, 15, 78
Euschoengastia latchmani, 150
— *ulcerofaciens*, 150
Eutrombicula species, 8, 10, 11, 38, 149
— *alfreddugesi*, 11, 38, 149
— *wichmanni*, 17, 28, 128, 131, 132, 138, 149
— —, confused, 39
evolution, relapsing fever, 114
—, trombiculids, 7
—, typhus fevers, 76
Ewing, H. E., 11, 28
experimental method, birthplace, 21
experimentalist and observer, 18
explosive outbreaks, 173
explorers and scrub-itch, 17, 128

Fabre, Henri, 19
fallacies, rat control, 117
famine, and bamboo flowering, 176
—, Irish potato, 64
feeding of chiggers, 15, 16
festivals, Japanese, 30, 171
fever, coastal, Queensland, 75
—, mosquito-borne, 40, 116
fevers, emergence of typhus, 63
—, non-specific, 17, 38
Field, J. W., 72
Field Typhus Research Team (India), 98
Figure, air photographs, Plate I (Fig 11), Plate II (Fig. 12)
—, cartoon, 179 (Fig. 13)
—, effigies of tsutsuga, 30 (Fig. 5), 31, 32 (Figs. 6, 7)
—, epidemiological, 6 (Fig. 4), Plates I, II (Figs. 11, 12)
—, *Leptotrombidium akamushi*, larva, i (Fig. 1)
—, life-cycle, Japanese, 3 (Fig. 2)
—, —, U.S. Army, 4 (Fig. 3)
—, —, epidemiological, 6 (Fig. 4)
—, map, foci in Japan, 42 (Fig. 10)
—, shrines, 33, 34 (Figs. 8, 9)
film, scrub-typhus, 100
Finschhafen, 174
fires, annual grass, 90
'first prejudice', 57, 96
Fletcher, William, 69
flea-typhus, 78
flies, biting, 145
Flood Fever, 29, 156, 169
food, of chiggers, 90
—, of rats, 94
Ford, W. K., 101
forest, gallery, 109, Plates I, II (Figs. 11, 12)
—, scrub-typhus in, 96, 106
Formosa, 53, 89, 171
Fort Dufferin, 174
freak laboratory accidents, 71, 74
fringe-habitats, 109
Fuda-nagashi ceremony, 35
Fuji, Mt., 42 (Fig. 10), 115
Fuller, Henry S., 99

INDEX

gallery forest, 109, Plates I, II (Figs. 11, 12)
gamble, in science, 58
Gan, Addu Atoll, 13
Garcia, Richard, 146
gardens, infestation, 12, 19, 108, 116
—, infection in, 116
Gater, B. A. R., 73
genetic changes, 123
geographic range, 12
Glossina, 146
Goodenough I., 174
Goodyear plantation, 69
gonone, 17, 128
grasses, 89, 92
grocer's itch, 148
Grosseteste, Robert, 21
Gunther, C. E. M., 104
gurud, 41

habitat and parasites, 12
haemorrhagic fever, mosquito-borne, 116
Haffner Publishing Co., viii, 6 (Fig. 4)
Hagurò festival, 30, 171 30 (Fig. 5)
hairy-mite, 28
Hales, Stephen, 19, 23
Harrison, J. L., 24, 94, 117, 120, 180, 179 (Fig. 13)
harvest-mite, 1, 2, 142
—, distribution, 10
—, rabbits and, 11, 18
Hashimoto, Hakuju, 44, 67
Hayakawa, Kiyoshi, 3 (Fig. 2)
Hatori, Juro, 29, 37, 53, 89, 171
head-hunters, 86
health, socio-cultural aspects, 22
Heaslip, W. G., 104
hedgerows, 109
Helenicula species, 10
Henderson, James, viii
Honzo Komoku (*Honsomo Mokusubu*) 37, 168
horses, influenced by flies, 145
hosts, behaviour and chiggers, 12, 13, 16

—, incidental and maintaining, 12, 6 (Fig. 4)
—, specificity, 13
human ecology, 22
humanities and sciences, 21, 22
hypothesis, mistaken, 44

identification of vectors, confusion of, 161
ideographs, *mushi* and *tsutsuga*, 36
immunity to mite, 17
Imphal, 86
— Circus, 85, 97
incidence, in sexes, 15, 16
—, occupational, 12, 13, 16
increasing endemicity, 101
incubation period, 17
India, distribution in, 78
—, routes to Burma, 85
infection, localization, 105
infestation, 'personal', of rats, 117
insects, vectors of typhus, 79
Institute for Medical Research, Malaya, viii, 69, 120, 174, 178, 179 (Fig. 13)
investigators, death by infection, 64
Irish famine, 64
Irrawaddy, 174
irrigation and encephalitis, 116
Ishii, Hitoshi, 171
'islands', distribution of, 28, 6 (Fig. 4), Plate II (Fig. 12)
islands, 99
itch, mites, 127
Ixodidae, 148

jail-fever, 2, 64
Jamshedpur, 108, 116
Japan, recent investigations, 52
— and scrub-typhus, 28, 42, 43, 78
Japanese chart, 3 (Fig. 2)
— personality, 58
— River Fever, 29, 66, 156, 157
— investigators, 170
Jarak I., 119
—, an oceanic island, 121
—, shooting rats, 118

INDEX

Jayewickreme, S. H., 90
jealousy, scientific, 55, 56
Jellison, William, 99
Jhingergacha, 87
Jizo, Kedani, 34
Joslyn, M. H., 178
'jungle tsutsugamushi', 95, 106
Jutland, scrub-itch, 12

Kabaw Valley, 86, 107, 109
Kakurai, Tokibumi, 35, 170
Kalewa, 86, 98, 173
Kalra, S. L., 98, 101
kaneo, for malaria, 40
Kanglatongbi, 102
Kawakami, Seiya, 39, 43, 45, 49
—, biography, 46
—, fate of, 49
Kawamura, Kiyoshi, viii, 29, 44, 53
kedani, 28, 39, 153, 156
Kedani-Daimyojin, shrine, 22, 35, 33 (Fig. 8), 154
Kedani Jizo, 34
Kedania tanakai, 39
Kepka, Otto, 47, 143
Keukenschrijver, N. C., 69
Kingsbury, A. N., 69
— strain, origin, 72
Kishida, Hisayoshi, 39
Kitaoka, Masami, viii, 30 (Fig. 5), 32
Kitasato Institute, 52
kleshchei, 76
kogon grass, 89
Kohima, battle of, 86
Kohls, Glen, 99, 117
Kono, shrine at, 35
kozui-netsu, 48, 157
krasnotelki, 76
Kuala Lumpur, 69
—, Institute at, viii, 69, 120, 174, 178, 179 (Fig. 13)
Kumada, Nobuo, vii, 29, 35, 170
kunai grass, 89

labels, ecological, 12
laboratory animal, search for, 74

—, British Army, research, 85
— infections, 169
—, Scrub-Typhus Research, Imphal, 74
—, U.S. Army, Myitkyina, 99, 105
Laelapidae, 148
Lake Victoria, 41
lalang grass, 89
Lawrence, T. J., 101
Ledo Pass, 85
leeches, 133, 136
leptospirosis, 68
Leptotrombidium species, 8, 10, 38, 39, 43, 52, 82
— akamushi, i (Fig. 1), 28, 38, 53, 82, 109, 114
—, confused identifications, 104, 161
— *deliense*, 14, 28, 87, 98, 109, 114, 121
— —, early record, 103
—, distribution, 109
—, turnover, 82
— *scutellaris*, 115
Lesslar, J. E., 70
Lewthwaite, R., v, vii, 73, 97, 179 (Fig. 13), 180
Ley, Herbert, 179 (Fig. 13)
life-cycle charts, 3, 4, 6 (Figs. 2, 3, 4)
—, harvest bug, 19
— -cycles of mites, 2
Lim Boo Liat, 121
Li Shih-chén (Li Shiting), 37, 168
livestock, influenced by pests, 146
localization of chiggers, 105
louse-typhus, 74, 77, 78
Lushai, famine, 176

Mackenzie, J. M. D., 103
McKechnie, D., 65, 72
McRae, Iris, viii
maintenance of parasites, 12
maintaining hosts, 6 (Fig. 4), 12
Madagascar, potential risk, 116
maladies, man-made, 25, 95, 97, 147
malaria and mosquito, same name, 40, 41
— transmission, 40, 41, 51

— and miasma, 165, 166
Malaysia, chiggers, 14
—, Institute for Medical Research, viii, 69, 120, 174, 178, 179 (Fig. 13)
—, scrub-typhus, 28, 69
Maldives, 98
maleo-nests, 130
Mandalay, 87, 174
Mandel, Margot, viii
Manipur, 86
man-made maladies, 25, 95, 97, 147
— malaria, 96
— waste land, 108
mange-mites, 127, 148
Map, foci in Japan, 42 (Fig. 10)
Marchette, Nyven, 76
mark-release experiments, 117, 119
mass-flowering bamboos, 176
Materia Medica, Chinese, 37
matlalzahuatl, 38, 159
Mawlu, 106
May, Jacques M., viii, 6 (Fig. 4)
medical geography, 14, 81
medicine, preventive, 26
Megapodius, 10, 17, 130
Megaw, J. W. D., 67, 75
Mehta, D. R., 90
Meiji era, 43
Meiktila, 87
Mellanby, Kenneth, 97
Mesostigmata, 76, 126
Metastigmata, 126
miasma hypotheses, 41, 49, 52, 165
Milne Bay, 174
Minai, Naila, viii, 29
Minami, Atsumi, viii
mite-burning, 31
mite-colony, -island, 5, 11, 12
mites, dermatoses, 148
Mite Hill, 87, 97
mite-repellents, 100
mite-typhus, 75
mites, 75, 127, 154
—, classification, 127
Mogami River, 33, 34 (Figs. 8, 9), 32, 42 (Fig. 10)
mokka, 17

molecular biology, 22, 123
moon, new, and sandflies, 131
Mooser, H., 74
mosquito and malaria, 40, 41, 51, 165, 166
Mossman fever, 66
Morris, J. N., 100
Mus musculus, 82
mushi, 36, 153, 155
—, derivation ideograph, 36
mushi-yaki, 31, 171, 172
Myitnge, 175
Myitkyina, 99, 105
myxomatosis, 15

Nadchatram, M., 149
Nagayo, M., 29
Nagino, Tadashi, 43, 44
names, derivation, 8, 153
natural balances, 91
nature, respect for, 93
Neoschoengastia gallinarum, 150
— *nuniezi*, 149
Neotrombicula autumnalis, 8, 11, 12, 15, 18, 142, 143, 149
— — Denmark, 12, 15, 18, 151
— —, rabbits, 15
— *desaleri*, 149
'new' diseases, 115
Newell, I. M., 148
New York, 76
Niigata Prefecture, 28, 44, 42 (Fig. 10)
nomenclature, rules of, 78
North America, chiggers, 11, 28
New Guinea, 17, 102, 129

observer and experimentalist, 18
obsession, biasing, 65
obstinacy, scientific, 106
occupations and risk, 12, 13, 16
Ogata, Norio, 29, 39, 49, 53, 54, 168
Ohtomo, Genkei, 30
oil palm estate, 72, 73
Omar Ahmad, Ungku, viii
Operation Tyburn, 98
Ornithodorus, 9, 40, 147
ornithosis, 77

INDEX

Otobius megnini, 146
outbreaks, and epidemics, 89
—, Bat I., 99
—, Calcutta, 87
—, Ceylon, 87
—, Embilipitiya, 87
—, Goodyear estate, 69, 89
—, Jarak I., 120
—, Jhingergacha, 87
—, Kuala Lumpur, 87
—, Mandalay, 87
—, Meiktila, 87
—, Mite Hill, 87
—, Myitnge, 175
—, oil palm estate, 72, 73
—, Palel, 101
—, pinpointing, 106
—, Ranchi, 87
—, Rangoon, 87
—, Senembah estate, 66, 88
—, Sumatra, 66, 68, 69, 88, 89
—, types of, 173
—, urban, 87
Oxford, experimental method, 20, 21

Palel, 86, 101
Palm, T. A., 45
pamphlet, Kakurai's, 35, 170
parang-vegetation, 108
parasites, adaptation, 13
— and habitat, 12
—, maintenance, 6 (Fig. 4), 12
—, patterns, 13, 14
Parkhurst, H. J., 149
patau, 157
patterns of disease, changing, 113
— of parasitization, 14
Perry, Commodore, 43
personality, Japanese, 58
personal attitude, 94
— infestation rats, 117
Pescadores Is., 16, 53, 108
pest-arthropods, 89, 145
pest-chiggers, 17, 145
Petrishcheva, P., 10, 147
Philip, C. B., vii, viii, 4, 29, 54, 78, 146, 174, 179 (Fig. 13)

physiology, 24
Pine Festival, 30
Pinwe, 106
pioneer communities, 92
'place' diseases, 119
plantations infected, 66, 69, 88, 89
'poisonous place', 40
population dynamics, rats, 121–3
— ecology, 24
— vectors across edges, 109
potato famine, 64
Poulsen, P. A., 18, 151
prayers against infection, 35
prejudices, scientific, 110
prevention, scrub-typhus, 178
—, relapsing fever, 41
preventive medicine, 22, 26
priority in rickettsial naming, 169
Prome, 174
Prostigmata, 126
protection, 100, *and see* shrines
Proteus bacillus, 71, 74
proxemics, 123
pseudotyphoid, 66, 68, 69
psittacosis, 77
public health, 26
punishment by ticks, 10, 147

Q fever, 77, 78
Queensland, 66, 75, 78, 116
quotations on scrub-itch, 128
— on scrub-typhus, 161

rabbits and harvest-mites, 11, 15, 18
Radford, C. D., 97, 101
Radovsky, F., 146
Ranchi, 87
Rangoon, 87
rat-bite fever, 74
rats, dangers of control, 117
—, encouragement of, 82, 88
— feeding on termites, 94
—, Jarak I., 120, 121
—, plagues and famine, 176
—, population dynamics, 121
Rattus bowersi, record of, 118

— in evolution, 81
— *norvegicus* and *rattus*, 82
rodent control, fallacies, 117
— -mites, 148
Rebstock, M. C., 180
recognition, early, of vectors, 40
record infestation, 118
red mites, Japan, 28
References—*see* ends of Chapters
—, pest arthropods, 143
—, — chiggers, 150
— — diptera, 146
— — ticks, 147
—, scrub-itch, 128, 150
—, scrub-typhus, 161
—, vector ecology, 161
—, voyagers' stories, 128
refugees from Burma, 86
Reiss-Gutfreund, R. J., 78 (Table 1)
relapsing fever, 40, 41, 55, 63, 114
— — and evolution, 114
— — same name as tick, 40
repeated occupation, 175
research, publications on war-time, 97, 161
— and rivalry, 54
Research Units, Kuala Lumpur, v
— —, Myitkyina, v, 99, 105
reservoirs, 5, 6 (Fig. 4)
Ribbands, Ronald, 100
Ricketts, H. T., 64
Rickettsia species, 5, 78 (Table I), 79
—, evolution, 76
— *orientalis*, 53
—*tsutsugamushi*, 53
Rickettsiae, 2, 4 (Fig. 3), 5, 76
rickettsialpox, 76, 78 (Table I), 116
Rijichin, 37, 168
rivalry, scientific, 54, 58
River Fever, Japanese, 29, 66, 156, 157
Rocky Mountain spotted fever, 29, 64, 65, 74, 78
Roonwal, M. L., 102
rotmilbe, 10
rouget, 2
Rudd, Robert, 57
Rudnick, Albert, 152

rural waste land, 108
Russian names, 76, 154

Sagae River, 33, 34
Sambon, Louis, 37, 148
sandflies, 130, 131
sand-mites, 38, 158
Sarcoptidae, 127, 148
Sasa, Manabu, 29, 163
Savoor, S. R., 73, 179 (Fig. 13)
scabies mites, 127, 148
Schoengastia species, 8, 17, 128, 149, 150
Schoengastiella ligula, 100
Schoutedenichia, 47
Schüffner, W., 66
science and humanities, 22
scientific obsession, 65
— obstinacy, 106
scrub-itch, 1, 8, 9, 15, 138, 147
— chiggers, 145, 149
—, confusion with infection, 38
—, Oudemans on, 128
—, publications on, 150
scrub-typhus, 2, 8, 28, 41, 72, 75, 78, Plate II (Fig. 12)
—, disappearance of, 30
—, earliest accounts, 29, 39
—, distribution, 42, 81, Plate II (Fig. 12)
— incidence, 86, 87
— in Japan, 43
— in Malaya, 69
—, names for, 153
—, occupational, 16, 42
— in Pescadores, 16
—, spread, 116
—, winter, 115, 158
Scrub-Typhus Research Laboratory, Imphal, 97, 99
seasons, chiggers and, 10
secondary vegetation, 109
Selborne, Natural History, 1, 19, 59
sensitivity to mites, 17, 18
settlement, advancing human, Plate I (Fig. 11)
serpent and *akamushi*, 35

INDEX

shashitsu, 37, 38, 158
Shichito fever, 115, 158
shifting cultivation, 88, 90, Plate I (Fig. 11)
shimamushi, 28, 38, 159
Shimokadoma, 34
ship fever, 64
shop typhus, 72
shrine, serpent god, 35
shrines to *akamushi*, 30, 32, 33 (Fig. 8), 34 (Fig. 9)
Shinano River, 42 (Fig. 10), 44
Simulium, 9, 145
Singapore, chigger-patterns, 13
—, Changi camp, 113
'slant', ecological, 23, 24
slash-and-burn cultivation, 88, 90, Plates I, II (Figs. 11, 12)
Smadel, Joseph E., v, 178, 179 (Fig. 13)
Smith, Marjorie, viii, 30 (Fig. 5)
Snake River Valley, 65
social use of space, 123
socio-cultural aspects of health, 22, 123, 124
soft ticks, 9, 146
soil-bacteria and rickettsiae, 80
Somaliland, names of vectors, 40
specificity, parasites, 13
speculation on evolution, 76
spinose ear-tick, 146
spirochaetoses, 40, 41, 55, 63, 114
Spooner, E. T. C., vii
sporadic cases, 174
spotted fevers, 63, 64, 67
Stanbury, William, 102
Stewart, H. C., 102
stress, in hosts, 14
students' attitudes, 22
stylostomes, 16
suison-netsu, 48, 159
Sumatra, 66, 68, 69, 88, 89
symptoms, 38
synecology, 24
System of Natural History, Chinese, 37
systems, 24

tabardillo, el, 63
Taiwan, 53, 89, 171
Tamiya, Takeo, v, viii, 30 (Fig. 5), 31 (Fig. 6), 32, 33, 34, 163
Tanaka, Keisuke, 29, 52
teaching, 20, 21, 22
termites, diet of rats, 94
terrain, types of, 108
test, Weil-Felix, 70
Tetranychidae, 148
Thisted, 12
Thomas, Harry M., 99
'tick', origin, 154
'tick sickness', 41
ticks, mistaken as vector, 102
—, punishment by, 10, 147
—, soft, 9, 40, 147
— and trombiculids unrelated, 80, 154
— -typhus, 64, 65, 67, 78 (Table I)
tlalzahuatl, 38, 159
Tomlin, D., 175
towns, scrub-typhus in, 87
tradition and bias, 68, 96
transmission, early ideas on, 40
— malaria, 40, 41, 165
— relapsing fever, 41
— transovarial, 5
Traub, Robert, 14, 99, 161, 179 (Fig. 13)
treatment of scrub-typhus, 178
tree-shrew, 103
Treftz, Julie, viii
trench fever, 78, 79
trombidiosis, 8, 147, 149
Trombicula, genus, 39
— *irritans*, 11
— *minor*, 39, 104
— species: see *Leptotrombidium*
tropical typhus, 69, 70, 72
tsetse flies, 146
tsutsuga, derivation, 36, 168
— effigies, 30–32 (Figs. 5–7), 172
tsutsugamushi, i (Fig. 1) *and see* chiggers, trombiculids
—, burning effigies, 30–32 (Figs. 5–7), 172

INDEX

—, burning hair, 46
—, disease, 3 (Fig. 2), 4 (Fig. 3), 28, 29, 38, 66, 74, 169 *and see* scrub-typhus
—, —, earliest accounts, 29
—, Japanese research, 41, 52, 54
—, 'jungle', 95, 106
tubes, drinking, 17
Tunga penetrans, 8
Tupaia belangeri, 103
Tuxen, S. L., 12
Tweedie, M. W. F., 121
typhoid fever, 63
typhus abdominalis, 63
— Commission, British, 97, 98
— —, U.S., 99, 105
—, endemic, 64, 65, 78 (Table I)
—, epidemic, 63, 64, 65, 74, 77, 78 (Table I)
— *exanthematicus*, 63
— fevers, forms of, 69, 78 (Table I)
— —, evolution, 76, 79
—, flea-, 63, 64, 78 (Table I)
— in Irish emigrants, 64
— group, emergence of, 63
— House, Mandalay, 174
— 'islands', 40, 88
—, louse-, 63, 65, 74, 77, 78 (Table I)
—, Malayan research, 75
—, *recurrens*, 63
—, scrub-, *see* scrub-typhus
—, tick-, 63, 64, 67, 78 (Table I)
—, treatment, 180
—, tropical, 69, 70
—, vectors, 78 (Table I)
typhoid, treatment, 180

Überschwemmungsfieber, 48, 160
Ukereri I., relapsing fever, 41
Ungku Dr Omar-Ahmad, viii
U.S. Army Medical Research Unit, Malaya, 179 (Fig. 13), 180
U.S. Typhus Commission, 99, 105

vaccine, cotton-rat, 98, 178
Valentine, Lucille, viii

van der Sande, G. A. J., 137
vanillism, 148
vectors, African names for, 40
—, arthropod, 40, 78 (Table I)
—, confusion of identities, 161
—, controversy about, 102
—, dispersal, 96
— and plant juices, 90
vegetation and scrub-typhus, 105
velvet-mites, 8, 28
Vercammen-Grandjean, P. H., i (Fig 1), 47, 142, 161
von Prowazek, C., 64

Wachsmuth, M., 66
Wallace, Alfred Russell, 17, 129
Walton, G. A., 41, 55, 147
waste land, 108
water-meadows, 108
Watkins, Rosalie, viii
weeds, 90, 92
—, animal, 25, 82, 91, 93
Weil-Felix test, 70
Weil's disease, 68
Western equine encephalitis, 116
Wetyu, 109
Wharton, G. W., 90, 151
White, Gilbert, 1, 10, 12, 15, 16, 18, 19, 59, 142
—, contrasted with others, 19
Wichmann, C. E. A., 136, 138
Wilson, W. G., 70
winter scrub-typhus, 115, 158
— chiggers, 10, 115
Wisseman, Charles D., 161
women and scrub-itch, 15
Wood, A. N., vii
Woodward, T. E., 179 (Fig. 13)
World War II, scrub-typhus, 53, 84, 160
Wyatt-Smith, John, 120

X. K. strain, origin of, 71

Yamagata Prefecture, 28, 30, 32, 35, 170
Yamaguchi, M., 53

Yamamiya, C., 53
yō, 36, 152
yochu-byo, 152, 160

yudokuchi, 40, 42, 88, 101, 173

zoonoses, 6 (Fig. 4), 119